# Geochemical Characteristics and Contamination Risk Assessment of Soil

# Geochemical Characteristics and Contamination Risk Assessment of Soil

Guest Editors

**Jasminka Alijagić**
**Srecko Stopic**

MDPI

Basel • Beijing • Wuhan • Barcelona • Belgrade • Novi Sad • Cluj • Manchester

*Guest Editors*

Jasminka Alijagić
Mineral Resources and Geochemistry
Geological Survey of Slovenia
Ljubljana
Slovenia

Srecko Stopic
Process Metallurgy and Metal Recycling
RWTH Aachen University
Aachen
Germany

*Editorial Office*
MDPI AG
Grosspeteranlage 5
4052 Basel, Switzerland

This is a reprint of the Special Issue, published open access by the journal *Minerals* (ISSN 2075-163X), freely accessible at: https://www.mdpi.com/journal/minerals/special_issues/7AY0XHK1IQ.

For citation purposes, cite each article independently as indicated on the article page online and as indicated below:

| Lastname, A.A.; Lastname, B.B. Article Title. *Journal Name* **Year**, *Volume Number*, Page Range. |
|---|

**ISBN 978-3-7258-5629-9 (Hbk)**
**ISBN 978-3-7258-5630-5 (PDF)**
**https://doi.org/10.3390/books978-3-7258-5630-5**

# Contents

# About the Editors

**Jasminka Alijagić**

Jasminka Alijagić is employed at the Geological Survey of Slovenia. She holds a PhD in Environmental Sciences from the University of Nova Gorica, Slovenia (2013). She has developed extensive expertise in evaluating the environmental impacts of mining and metallurgical activities on soils and river sediments. Her expertise includes modelling using linear and non-linear mathematical methods. Additionally, I have gained knowledge and experience in mapping potential mineral resources, assessing secondary raw materials (SRM), and developing environmentally sound processes for recovering valuable materials from SRM. My advanced research and analytical skills enable me to tackle complex environmental challenges. She was leading and participating in many national and international projects.

**Srecko Stopic**

Srecko Stopic was born on April 3, 1965. in Uzice, Serbia. where he completed primary and secondary school. He graduated in 1991, master's degree in 1994 and doctorate in 1997 at the Faculty of Technology and Metallurgy in Belgrade, Serbia, majoring in non-ferrous metallurgy. In 2001, he received the Alexander von Humboldt research scholarship for Scientific Excellence and transferred to the RWTH Aachen University in Germany, where he leads projects in the field of extractive metallurgy at the Institute for Process Metallurgy and Metal Recycling (EURO-TITAN). He holds courses in extractive metallurgy of rare earth elements and Hydrometallurgy and electrolysis at the RWTH Aachen University.

minerals

MDPI

Editorial

# Geochemical Characteristics and Contamination Risk Assessment of Soil

**Srecko Stopic [1,*] and Jasminka Alijagić [2]**

[1] IME Process Metallurgy and Metal Recycling, RWTH Aachen University, Intzestrasse 3, 52056 Aachen, Germany
[2] Geological Survey of Slovenia, Dimičeva Ul. 14, 1000 Ljubljana, Slovenia; jasminka.alijagic@geo-zs.si
* Correspondence: sstopic@ime-aachen.de; Tel.: +49-17678261674

## 1. Introduction

Soil, as a key component of the four circles interconnecting atmosphere, hydrosphere, biosphere, and lithosphere, plays a vital role in sustaining human life and the terrestrial system globally. Meanwhile, soil contamination of toxic trace metals such as mercury is becoming an increasingly serious issue in many countries around the world, along with urbanization and industrialization, both of which pose severe risks to ecosystems and human health [1]. An increased content of trace metals can adversely affect the biological properties of soil, cause changes in the food chain, have a toxic effect on plants, and contaminate groundwater. Measuring the total concentration and geochemical baseline concentration are usually the first steps in the assessment of trace metal pollution [2]. Moreover, the background values can distinguish between natural and anthropogenically influenced concentrations, and through these values, in turn, one can calculate the anthropogenic contribution rate. When the permissible content level is exceeded, heavy metals reduce soil fertility, inhibit soil enzymatic activity, and change soil acidity.

Traditional long-lasting mining and extraction processes are accompanied by severe environmental pollution and ecological damage globally, but at the same time, the mining and mineral extraction processes industry plays a vital role in the development of modern technologies. There is a growing demand for raw material production and supplying invaluable resources for modern life. Modern society achieves greater efficiency by developing innovative waste recycling technologies [3].

The increase in the amount of mining and metallurgical waste represents a big problem for the environment. This is why various methods are used for the processing of flotation tailings, red mud, and slag landfills, which are a potential source of accumulated metals [4]. Unfortunately, they pile up on the ground and create an intractable problem. This is why studying the processing of various landfills is a great challenge affecting the scientific community and, thus, the entirety of humanity [5].

This article provides an overview of advances in understanding the importance of geochemical characteristics and soil contamination risk assessment while also covering new challenges regarding sustainable treatment of bauxite residue and the recovery of valuable elements. In particular, the reduction of iron oxides from bauxite residue before the leaching process can facilitate a decrease in the quantity of disposed material and soil pollution.

## 2. Contents of This Special Issue

This Special Issue contains eight papers, listed at the end of this article. Focusing on bauxite residues, problems created by residues of metallurgical production and the ways

 https://doi.org/10.3390/min15101011

in which these problems can be solved via the valorization of critical metals are explained. Understanding the temporal–spatial distribution and influencing factors of heavy metals on a regional scale is crucial for assessing the anthropogenic impacts and natural variations in elemental geochemical behavior. The study by Tian et al. (Contribution 1) evaluated the spatial distributions of heavy metals such as As, Cd, Pb, and Zn, as well as their driving mechanisms, over the past 31 years in Guangxi, China, using three geochemical baseline projects (the Environmental Geochemical Monitoring Network Project (EGMON) project (1992–1996); the Geochemical Baseline (CGB) 1 project (2008–2012); and the CGB2 project (2015–2019)). By calculating the variable importance using the random forest algorithm, it was found that natural factors were the primary drivers of the spatial distribution of heavy metals in the EGMON project, especially for the precipitation of As, while the digital elevation model (DEM) was the primary driver for Cd and Pb, and temperature was the primary driver for Zn. Surface alluvial soils showed obvious heavy metal enrichment in the CGB1 project, with the gross domestic product (GDP) driving the spatial distribution of all heavy metals. In addition, the anomalous intensity and range of heavy metals in the CGB2 project decreased significantly compared to the CGB1 project, likely due to the normalized difference vegetation index (NDVI) serving as a positive anthropogenic factor that improves the degree of rocky desertification, thus reducing the heavy metal contents of As and Pb, and precipitation promoted the decomposition of Fe–Mn concretions and, thus, the migration of Cd and Zn.

The features of structural and phase transformations during the processing of alloyed metallurgical wastes using reduction smelting are discussed herein. This is necessary in order to determine the technological parameters of the melting process that ensure reduction in the losses of alloying components. The use of X-ray phase analysis in combination with the methods of raster electron microscopy and X-ray microanalysis ensured the identification of the microstructure features and the chemical composition of individual phases and inclusions in the metal. The study by Smirnov et al. (Contribution 2) identified new technological aspects of high-alloyed technogenic waste processing using reduction smelting. The obtained parameters of the resource-saving alloying compound make it possible to replace parts of the standard ferroalloys in steelmaking processes.

The study by Silin et al. (Contribution 3) includes a combined hydrometallurgical treatment of an eudialyte ore sample with subsequent preparation of leaching residue using mechanical separation methods. Hydrometallurgical treatment contains dry digestion with sulfuric acid at room temperature and filtration of the obtained product. The objectives of adopting these procedures are to test a new digestion reactor in order to prevent silica gel formation from the eudialyte ore. The obtained results revealed that silica gel formation is prevented during dissolution with sulfuric acid. A high leaching efficiency of light rare earth elements (La, Ce, Nd, and Y) was reached using the dry digestion process with sulfuric acid, where the starting molarity was 12 mol/L. After the filtration process, magnetic separation was studied as the main method to recover weakly magnetic minerals like amphiboles and pyroxenes from the leaching residue in the magnetic fraction and feldspars in the nonmagnetic fraction. A new combined research strategy was developed for the production of different concentrates such as the one bearing Zr, Hf, and Nb.

The more economically viable and environmentally sustainable approach for treating the by-products of coal combustion from thermal power plants entails their collective disposal as opposed to individual disposal methods. This aligns with pertinent EU directives and domestic regulations, ensuring compliance with established standards while optimizing resource utilization and minimizing environmental impact. This study evaluated the resistance to wind erosion of the binding properties of a mixture (comprising fly ash (FA), bottom ash (BA), and additives) using an indoor wind tunnel under simulated ambient conditions. Investigations of the mutual impact of ash, bottom ash, and additives (CaO and $Ca(OH)_2$) with a certain percentage of water were carried out with eighteen samples. The samples consisted of water at six addition rates 5, 8, 10, 15, 20, and 25% (*w*/*w*) and an additive at three addition rates (1, 2, and 3% (*w*/*w*)). Based on the obtained results, the optimal ratios of the additives (3% (*w*/*w*)) and water (15% (*w*/*w*)) were determined. Prior to the wind tunnel experiments, and according to the different addition rates of additives and water, eight samples were prepared with different addition rates of ash. The mass concentrations of suspended particles ($PM_{10}$) and total suspended particles (TSPs) in these samples were measured at three distinct wind velocities, 1 m/s, 3 m/s, and 5 m/s, respectively. The results reported by Petkovic-Papalazarou et al. (Contribution 4) indicate that the samples containing the optimal content of additives and water demonstrate a maximum increase in $PM_{10}$ emission zero values of no more than 1.9 times. This finding can be considered satisfactory from the standpoint of environmental protection.

The study by Lahori et al. (Contribution 5) holds significant implications. It investigates the comparative effect of biochar, zeolite, and bentonite minerals on the stabilization of Ni fractions, bioaccumulation, translocation indices, and the reduction in their absorption by pakchoi in smelter- and mine-contaminated soils. The results, which are of great interest, show that the maximum fresh and dry biomasses of pakchoi were 28.21 and 18.43% for smelter-polluted soil and 61.96 and 67.90% for mine-contaminated soil amended with zeolite compared to the control. Applying zeolite increased pakchoi chlorophyll SPAD values 1.17-fold in smelter soil and 1.26-fold in mine-polluted soil. The highest level of Ni immobilization in the smelter and mine soil was 76.8 and 85.38% with the application of bentonite, which increased soil pH and CEC. The application of biochar, bentonite, and zeolite reduced the Ni residual, oxidizable, and acid-soluble fractions, but biochar and bentonite increased the reducible fraction of Ni in smelter soil. The highest reduction in Ni in the shoot and root was noted as 82.08 and 68.28% of smelter-polluted soil and 77.25 and 89.61% of mine-polluted soil with bentonite compared to the control soil. Overall, it has been concluded that biochar, zeolite, and bentonite can be successfully used to mitigate the Ni concentration in smelter- and mine-polluted soil and reduce uptake by vegetable crops.

The global generation of bauxite residue necessitates environmentally responsible disposal strategies. The included study by Mendy et al. (Contribution 6) investigated the long-term (5-year) behavior of bauxite residue with pH lowered to 8.5, called modified bauxite residue (MBR), using lysimeters to test various configurations—raw MBR or used MBR (UMBR) previously applied for acid mine drainage remediation, sand or soil capping, and revegetation. Throughout the experiment and across all configurations, the pH of the leachates stabilized between 7 and 8, and their salinity decreased. Their low sodium absorption ratio (SAR) indicated minimal risk of material clogging and suitability for salt-tolerant plant growth. Leaching of potentially toxic elements, except vanadium, decreased rapidly after the first year to low levels. Leachate concentrations consistently remained below LD50 for *Hyalella azteca* and were at least an order of magnitude lower by the experiment's end, except for first-year chromium. Sand capping performed poorly, while revegetation and soil capping slightly increased leaching, though these increases were negligible, given the low final leaching levels. Revegetated MBR shows promise as a suitable and sustainable solution for managing bauxite residues, provided the pH is maintained above 6.5. This study highlights the importance of long-term assessments and appropriate management strategies for bauxite residue disposal.

Assessment of the plant's ability to take up mercury (Hg) from polluted soil was affected by location, plant family, and species in two former cinnabar mining areas in the Czech Republic. At each location, seven sampling points were marked out in the vicinity of former shafts and dumpsites connected to the mining activity, where representative soil samples and dicotyledonous plants were collected. The individual locations were characterized by specific plant communities, where, in most cases, different plant species were found within one family at both locations. The total Hg content in the soil, as well as gaseous elemental mercury ($GEM_{soil\text{-}air}$), confirmed elevated levels of this element in the mining-affected environment, along with high data variability. The low Hg accumulation ability of plants, especially the low root–shoot translocation in most plant species, indicated the predominant occurrence of excluders. The results reported by Bauštein et al. (Contribution 7) showed the exceptional position of the Fabaceae family regarding soil Hg pollution, as the highest Hg content in both shoots and roots was determined for *Onobrychis viciifolia*. Therefore, the behavior of Fabaceae plants in polluted soil, the mechanisms of their tolerance to high Hg content, and their Hg accumulation ability deserve further research.

Comparative analysis of red mud reduction techniques was performed by Stopic et al. (Contribution 8) using both carbothermal and hydrogen-based reduction methods, combining thermochemical modeling and experimental validation. The reduction process is important because of the soil contamination risk assessment with disposed red mud. Therefore, the minimization of red mud during the reduction process could be a novel strategy for the production of metallic iron and solid residue for hydrometallurgical treatment. Different strategies of hydrogen and carbon reduction in static and dynamic conditions were studied between 700 °C and 1700 °C. The separation of solid residue and formed iron was analyzed using magnetic separation. The main aim was to study the advantages and disadvantages of using decarbonizing technologies for the treatment of red mud, aiming to develop an environmentally friendly process. Thermochemical analysis of the reduction offered new data about mass losses during our process through evaporation, thermal decomposition, and the formation of metallic carbide.

## 3. Conclusions

Five years of leaching experiments to evaluate land spreading of a modified bauxite residue before and after acid mine drainage treatment with comparative analysis of reduction techniques, aimed at the minimization of contaminated soil with bauxite residues, represent the recent advances in understanding soil contamination risk assessment. Assessment of mercury uptake by plants in former cinnabar mining areas also represents an additional contribution. Prevention of silica gel formation for eudialyte study using a new digestion reactor can be achieved during bauxite residue treatment. The comparative effect of sustainable materials on the immobilization, geochemical fractions, bioaccumulation, and translocation of Ni in smelter- and mine-polluted soils has also been studied. The effects of wind velocity on the binding properties of ash, bottom ash, and additives were studied via a wind tunnel study. The temporal–spatial distributions and influencing factors of certain heavy metals—namely As, Cd, Pb, and Zn—in alluvial soils in Guangxi, China, have also been studied. In addition, it has been found that recycling of alloyed metallurgical waste depends on structure and phase transformation.

Ultimately, this Special Issue, titled "Geochemical Characteristics and Contamination Risk Assessment of Soil", comprises eight important papers which offer new insights and results needed for soil contamination risk assessment in the context of geochemical characteristics and metallurgical and mining activities, which expose waste materials. This is a warning to the wider community to become involved in finding solutions to these problems by considering the latest results. Unfortunately, the Editors of this Special Issue do not have high expectations regarding the implementation of the included insights, because all planned actions always come with large costs and problems with securing the necessary funding.

**Conflicts of Interest:** The authors declare no conflicts of interest.

**List of Contributions:**

1. Tian, M.; Wang, X.; Zhou, J.; Chi, Q.; Liu, D.; Qiao, Y.; Wu, C.; Zhou, Y.; Wang, Q.; Liu, F. Temporal–Spatial Distributions and Influencing Factors of Heavy Metals As, Cd, Pb, and Zn in Alluvial Soils on a Regional Scale in Guangxi, China. *Minerals* **2023**, *13*, 1107. https://doi.org/10.3390/min13081107.
2. Smirnov, A.; Petryshchev, A.; Bilko, T.; Andreev, A.; Semenko, A.; Skorobagatko, Y. Development of the Recycling of Alloyed Metallurgical Waste: Features of Phase and Structural Transformations. *Minerals* **2023**, *13*, 1171. https://doi.org/10.3390/min13091171.
3. Silin, I.; Dertmann, C.; Cvetković, V.S.; Stopic, S.; Friedrich, B. Prevention of Silica Gel Formation for Eudialyte Study Using New Digestion Reactor. *Minerals* **2024**, *14*, 124. https://doi.org/10.3390/min14020124.
4. Petković Papalazarou, S.; Nešković, J.; Ćorluka, S.; Polavder, S.; Mitrašinović, A.; Stjepanović, P. The Effects of Wind Velocity on the Binding Properties of Ash, Bottom Ash, and Additives: A Wind Tunnel Study. *Minerals* **2024**, *14*, 809. https://doi.org/10.3390/min14080809.
5. Lahori, A.H.; Ahmed, S.R.; Afzal, A.; Mierzwa-Herstek, M.; Umar, A.; Muhammad, M.T.; Vambol, V.; Saleem, I.; Tanoli, A.A.; Vambol, S. A Comparative Effect of Sustainable Materials on the Immobilization, Geochemical Fractions, Bioaccumulation, and Translocation of Ni in Smelter- and Mine-Polluted Soils. *Minerals* **2024**, *14*, 1083. https://doi.org/10.3390/min14111083.
6. Merdy, P.; Parker, A.; Chen, C.; Hennebert, P. Five Years of Leaching Experiments to Evaluate Land Spreading of a Modified Bauxite Residue Before and After Treatment of Acid Mine Drainage: Sand or Soil Capping and Revegetation. *Minerals* **2024**, *14*, 1210. https://doi.org/10.3390/min14121210.
7. Bauštein, M.; Száková, J.; Stefanović, L.; Najmanová, J.; Sysalová, J.; Tlustoš, P. Assessment of Mercury Uptake by Plants in Former Cinnabar Mining Areas. *Minerals* **2024**, *14*, 1211. https://doi.org/10.3390/min14121211.
8. Stopić, S.; Kostić, D.; Perušić, M.; Schneider, R.; Souza Filho, I.R.; Mitrašinović, A.; Friedrich, B. Comparative Analysis of Reduction Techniques Aiming for the Minimization of Contaminated Soil with Red Mud. *Minerals* **2025**, *15*, 470. https://doi.org/10.3390/min15050470.

## References


1. Gworek, B.; Dmuchowski, W.; Baczewska-Dąbrowska, A.H. Mercury in the terrestrial environment: A review. *Environ. Sci. Eur.* **2020**, *32*, 128. [CrossRef]
2. Peralta-Videa, J.R.; Lopez, M.L.; Narayan, M.; Saupe, G.; Gardea-Torresdey, J. The biochemistry of environmental heavy metal uptake by plants: Implications for the food chain. *Int. J. Biochem. Cell Biol.* **2009**, *41*, 1665–1677. [CrossRef] [PubMed]
3. Balomenos, E.; Gianopoulou, I.; Panias, D.; Paspaliaris, I. ENEXAL: Novel technologies for enhanced energy and exergy efficiencies in primary aluminium production industry. *Metall. Mater. Eng.* **2009**, *15*, 203–217.
4. Cheema, H.A.; Ilyas, S.; Farhan, M.; Yang, J.-H.; Kim, H. Extraction of Fe and Al from Red Mud Using Carbothermic Reduction Followed by Water Leaching. *J. Ind. Eng. Chem.* **2024**, *136*, 201–210. [CrossRef]
5. Hedenquist, J.W.; Antonio Arribas, R.; Gonzalez-Urien, E. *Exploration for Epithermal Gold Deposits, Gold in 2020, Society of Economic Geologists*; GeoScienceWorld: McLean, VA, USA, 2000; Volume 13, ISBN 9781629490205. [CrossRef]

*minerals*

MDPI

*Article*

# Temporal–Spatial Distributions and Influencing Factors of Heavy Metals As, Cd, Pb, and Zn in Alluvial Soils on a Regional Scale in Guangxi, China

**Mi Tian [1,2,*], Xueqiu Wang [1,2,*], Jian Zhou [1,2], Qinghua Chi [1,2], Dongsheng Liu [1,2], Yu Qiao [3], Chao Wu [1,2], Yining Zhou [1,2], Qiang Wang [1,2] and Futian Liu [4]**

1 Key Laboratory of Geochemical Exploration, Institute of Geophysical and Geochemical Exploration, Langfang 065000, China
2 International Centre on Global-Scale Geochemistry, Langfang 065000, China
3 Zijin Mining Group Co. Ltd., Longyan 364200, China
4 Key Laboratory of Strategic Mineral Resources of the Upper Yellow River, Ministry of Natural Resources & School of Earth Sciences, Lanzhou University, Lanzhou 730000, China
* Correspondence: tianmi62080608@126.com (M.T.); wxueqiu@mail.gov.cgs.cn (X.W.)

**Abstract:** Understanding the temporal–spatial distribution and influencing factors of heavy metals on a regional scale is crucial for assessing the anthropogenic impacts and natural variations in elemental geochemical behavior. This study evaluated the spatial distributions of the heavy metals As, Cd, Pb, and Zn as well as the driving mechanisms over the past 31 years in Guangxi, China, using three geochemical baseline projects (the Environmental Geochemical Monitoring Network Project (EGMON) project 1992–1996; the Geochemical Baseline (CGB) 1 project 2008–2012; and the CGB2 project 2015–2019). By calculating the variable importance using the random forest algorithm, it was found that natural factors are the primary drivers of the spatial distribution of heavy metals in the EGMON project, especially precipitation for As, the digital elevation model (DEM) for Cd and Pb, and temperature for Zn. Surface alluvial soils showed obvious heavy metal enrichment in the CGB1 project, with the gross domestic product (GDP) driving the spatial distribution of all heavy metals. In addition, the anomalous intensity and range of heavy metals in the CGB2 project decreased significantly compared with the CGB1 project, especially owing to the normalized difference vegetation index (NDVI) as a positive anthropogenic factor that improves the degree of rocky desertification, thus reducing the heavy metal contents of As and Pb, and the precipitation promoting the decomposition of Fe–Mn concretions and thus the migration of Cd and Zn. This research promotes an understanding of anthropogenic and natural influences on the spatiotemporal distribution of heavy metals and is of great significance for environmental monitoring and governance.

**Keywords:** heavy metals; alluvial soil; spatiotemporal variations; driving factors; geochemical baseline projects

## 1. Introduction

Heavy metals are harmful pollutants that are covert, persistent, irreversible, and easily enter the human body through the food chain, causing damage and pathological changes in organs and thus threatening human health [1–5]. Heavy metals are widely recognized as a global environmental threat [6–8]. Therefore, it is of great significance to study the spatiotemporal distributions of heavy metals to monitor anthropogenic disturbances and naturally occurring changes. However, previous studies on the spatiotemporal distributions of heavy metals have mostly been based on data from two periods and have mostly been conducted at the local scale [9–11]. Moreover, previous studies have mainly focused on spatial–temporal change characteristics, ignoring the quantitative evaluation of the main driving factors of spatiotemporal changes [12,13].

China has conducted three environmental geochemical baseline projects aimed at monitoring the changes in concentration and spatial distribution of geochemical elements caused by anthropogenic activities, climate change, and geological processes, thus providing a quantitative scale for environmental change, including the Environmental Geochemical Monitoring Network Project in 1992–1996 (EGMON) [14,15] and China's first and second Geochemical Baseline projects in 2008–2012 (CGB1) [16] and 2015–2019 (CGB2) [17]. These projects aimed to provide high-resolution and high-quality geochemical baseline data by developing improved sampling and laboratory analysis methodologies. The primary objective of the projects was to establish nationwide geochemical baselines against which future human-induced or natural chemical changes could be recognized and quantified. Samples of alluvial soil, the most representative sampling medium, which represents the average values of elements and effectively reflects the environmental changes caused by human input in the basin, were collected. Alluvial soils can be used to monitor environmental changes and inputs because the transport processes of pollutants in rivers occur continuously [18]. The similar sample media, analytical methods, and quality controls used by the three projects may provide an unbiased opportunity to quantify these changes [19].

The driving factors of spatiotemporal variations in heavy metals generally include external inputs; the influence of parent materials; and geochemical processes, such as leaching and transformation through surface runoff [20,21]. Heavy metals in soils are sourced from parent materials, the concentrations of which are usually increased by external inputs, whereas both are increased or decreased by geochemical processes, depending on the different elements and environmental conditions [22]. Therefore, identifying the main driving factors and quantifying their influence on the spatiotemporal variations of heavy metals is helpful in determining the sources of heavy metals and providing meaningful information on anthropogenic disturbances and naturally occurring changes [23–25].

Random forest is a highly accurate, adaptable, and interpretable machine learning method that uses a set of decision trees to classify and regress and can incorporate both continuous variables and type variables simultaneously [26]. Moreover, the predictive values of the dependent variables and the relative importance of each variable can be calculated on the basis of the nonlinear relationships between the dependent and independent variables. The random forest method is becoming increasingly popular in many fields, such as geological mapping [27], digital soil mapping [28], and mineral exploration mapping [29].

This study investigated the spatiotemporal distributions and variations of the heavy metals As, Cd, Pb, and Zn in Guangxi, China, a typical province with a widely distributed karst area, and quantified the driving factors impacting heavy metal concentrations at regional scales using a random forest algorithm based on data obtained from three geochemical baseline projects. This research is helpful for understanding the geochemical behavior of heavy metals and quantitatively evaluating anthropogenic disturbances and natural changes over a long period of time.

## 2. Materials and Methods

### 2.1. Study Area

The total area of Guangxi province, located in South China (104°26′ E~112°04′ E; 20°54′ N~26°24′ N), is about 23.67 × $10^4$ km$^2$, and the province has a subtropical monsoon humid climate (Figure 1). The terrain inclines from the northwest to the southeast and is dominated by basins. Karst landforms are widely distributed in central, western, and northwestern Guangxi, covering an area of approximately 9.58 × $10^4$ km$^2$, accounting for 40.9% of the total area (Figure 2).

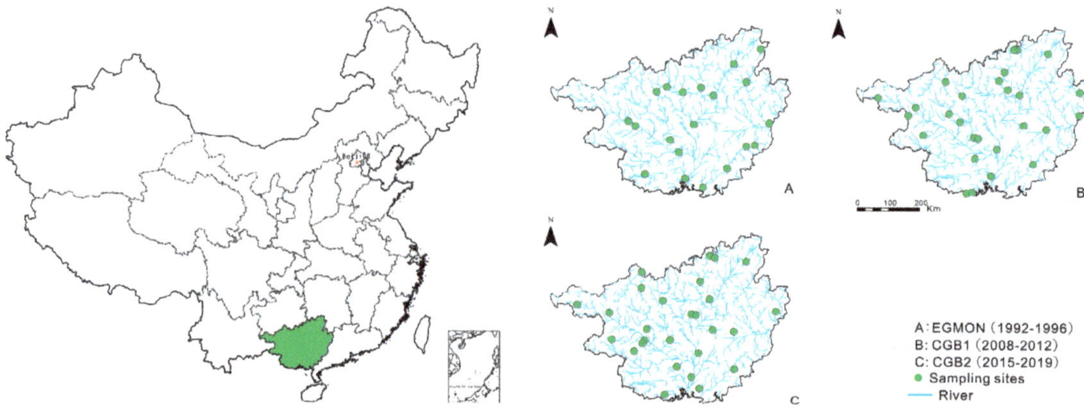


**Figure 1.** Map showing locations of the study area and samples taken by EGMON, CGB1, and CGB2. EGMON: Environmental Geochemical Monitoring Network Project, CGB: Geochemical Baseline project.

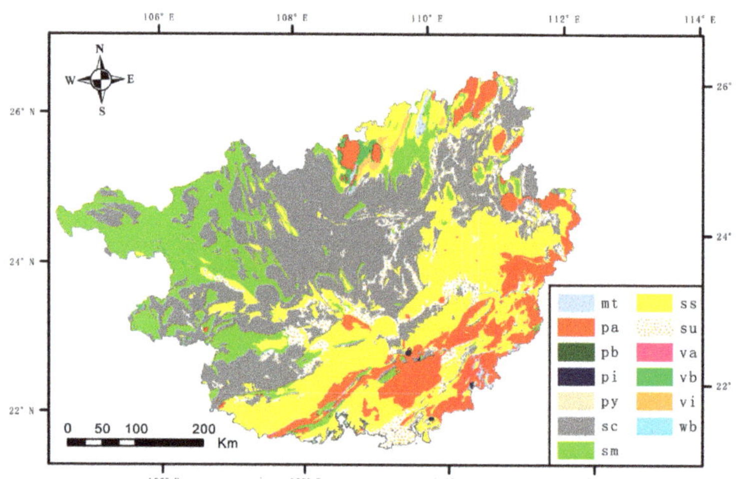


**Figure 2.** Guangxi geological map. mt: metamorphic rocks, pa: acidic plutonic rocks, pb: basic plutonic rocks, pi: intermediate plutonic rocks, py: pyroclastic, sc: carbonate sedimentary rocks, sm: mixed sedimentary rocks, ss: siliciclastic sedimentary rocks, su: unconsolidated sediment, va: acidic volcanic rocks, vb: basic volcanic rocks, vi: intermediate volcanic rocks, wb: water body.

### 2.2. Sampling and Analysis Method

#### 2.2.1. Materials and Sampling

In view of recognizing the need for global-scale geochemical baselines that can quantify future human-induced or natural changes in the chemistry of the Earth based on Global Reference Network grid sampling of Earth's surficial materials [30], China initiated the Environmental Geochemical Monitoring Networks (EGMON) project from 1992 to 1996 [14]. Most of the samples were collected in 1995. Samples of Alluvial soil, formed by flood sediments, were collected, and the sampling locations were mostly in the floodplains of large catchment basins ranging from 1000 to 10,000 km$^2$ [15]. The first China Geochemical Baselines Project (CGB1), as part of the Global Geochemical Baselines Project [30], was conducted from 2008 to 2014 [17], and most samples were collected in 2010–2012. The Second China Geochemical Baselines Project (CGB2) was carried out between 2015 and

2019 [18]. Samples of alluvial soil, formed from drainage sediments, were collected from 3382 and 1741 locations in CGB1 and CGB2, respectively, according to a global reference network grid cell in mainland China (9.6 million km$^2$). Sample locations were designated at the outlet plains of drainage catchments ranging in area from approximately 1000 to 5000 km$^2$, with most locations being 2000–3000 km$^2$ in area. The topsoil samples were collected from the surface to a depth of 25 cm, and the litter was scraped off; the minimum weight of each sample was 2.5 kg.

Temporal changes in heavy metal concentrations may increase or decrease, and a major issue is whether these changes can be detected by soil monitoring. Sampling materials and monitoring sites must indicate that contaminants build quickly enough to be revisited on subsequent occasions [31]. The sample media from the EGMON and CGB were the same; both were alluvial soils formed from catchment sediments (overbank/floodplain/delta sediments) through river transportation (Figure 3). All runoff materials are transported to the same outlet or plain to form soil through drainage network channels (Figure 3). The transported samples collected from the outlets of large drainage catchments are excellent media, representing the natural background and anthropogenic emissions of the area. The contamination of alluvial soils occurs relatively quickly. Pollution comes from diffuse sources, such as natural weathering, mining, industries, residents, pesticides, and fertilizers. Rainfall on land picks up and transports pollutants into watercourses and deposits them in low-reach plains, overbanks, or fluvial terraces.

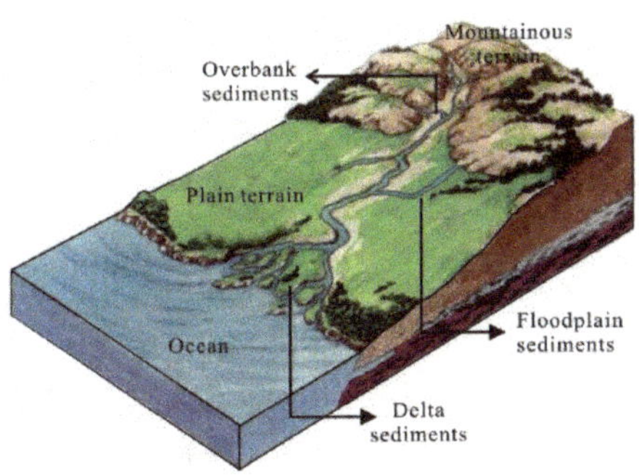


**Figure 3.** Figure showing the sampling sites.

Twenty samples from the EGMON project and 26 from the CGB1 and CGB2 projects were collected in Guangxi, as shown in Figure 1.

2.2.2. Sample Preparation and Laboratory Analysis

All samples were prepared and subjected to chemical analyses in the same laboratory. The samples were air-dried and homogenized, and each raw sample was split into two sub-samples; one was sieved through a mesh (<2 mm) for laboratory analysis, and the other was stored for future investigation. The sieved sample was ground to <74 μm in an agate mill for laboratory analysis.

An aliquot (0.25 g) was weighed and placed in a test tube, and 10 mL of HF, 5 mL of $HNO_3$, and 2 mL of $HClO_4$ were added to digest the samples. The test tube was heated in a boiling water bath until it dried. After cooling, 8 mL of 1:1 aqua regia (aqua regia (1 $HNO_3$ + 3 HCl): pure water = 1:1 vol.) was added to decompose the residue. The solution was diluted with 2% $HNO_3$ and then analyzed by ICP-MS to determine Cd and Zn. Arsenic was determined by hydride atomic fluorescence spectrometry, and Pb was determined by X-ray fluorescence spectrometry [32]. The detection limits were 1.00 mg/kg, 0.02 mg/kg, 2.00 mg/kg, and 4.00 mg/kg for As, Cd, Pb, and Zn, respectively. The accuracy of the method was assessed by analyzing the soil reference materials (GSS-1, GSS-2, GSS-17, GSS-19, GSS-25, GSS-26, GSS-27) [33] 34 times, and the ΔlgC was less than

0.10 ($\Delta \lg C = |\lg C_i - \lg C_s|$; $C_i$ is the average of measured values and $C_s$ is the standard reference value).

## 2.3. Factor Selection

Four anthropogenic factors—land use, the spatial distribution density of major pollution sources, gross domestic product (GDP), and the normalized difference vegetation index (NDVI) in Guangxi—were selected, and the geology background, temperature, precipitation, and digital elevation model (DEM) were selected as natural factors. Geology background and land use were type variables, whereas the others were numeric variables. Gross domestic product, NDVI, and land use were variable factors, and data from 1995, 2010, and 2015 were selected to represent the factor characteristics of the EGMON, CGB1, and CGB2 sampling periods. Data on land use, GDP, NDVI, temperature, and precipitation were collected from the Resource Environmental Science and Data Centre, Institute of Geographical Sciences and Natural Resources Research, Chinese Academy of Sciences (https://www.resdc.cn/data.aspx?DATAID=123, accessed on 17 January 2022). Temperature and precipitation data are presented as annual mean values. Digital elevation model data were collected from the Geospatial Data Cloud Website (http://www.gscloud.cn/search, accessed on 26 January, 2021), and major pollution source data for Guangxi were collected from the Guangxi Hechi Ecological Environment Bureau (http://sthjj.hechi.gov.cn/, accessed on 26 January, 2021). The spatial analysis module of ArcGIS 10.4 was used to calculate the nuclear density to obtain the pollution source density of Guangxi. The spatial distributions of the influencing factors in Guangxi Province are shown in Figures 4–7.

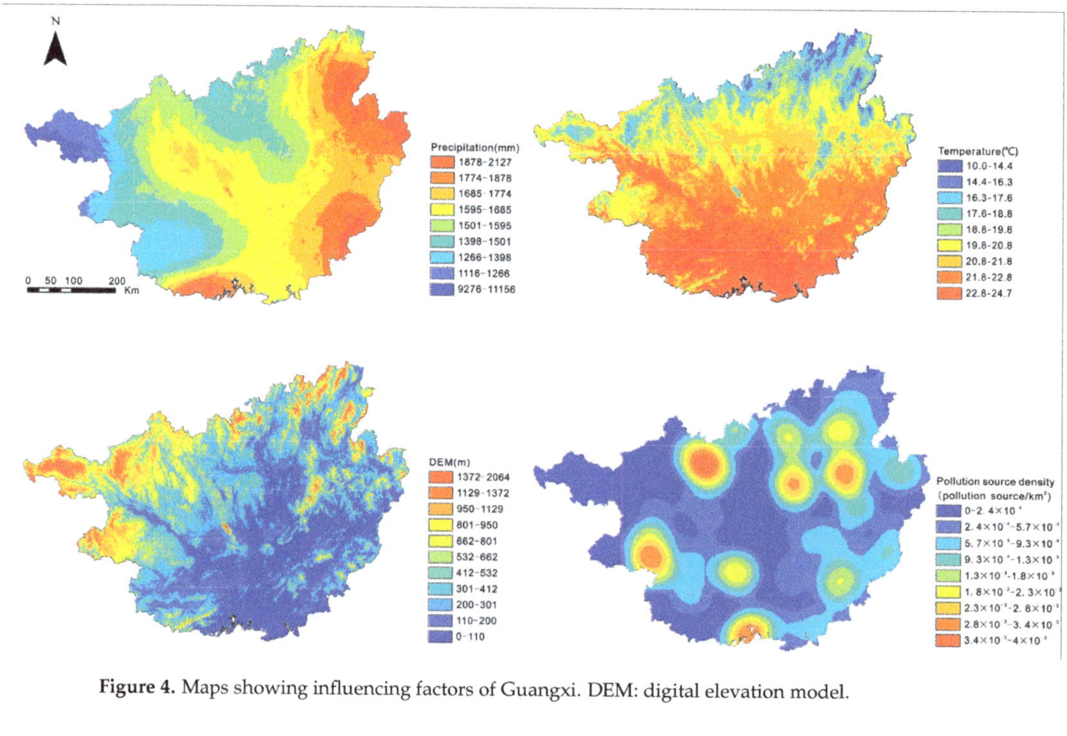


**Figure 4.** Maps showing influencing factors of Guangxi. DEM: digital elevation model.

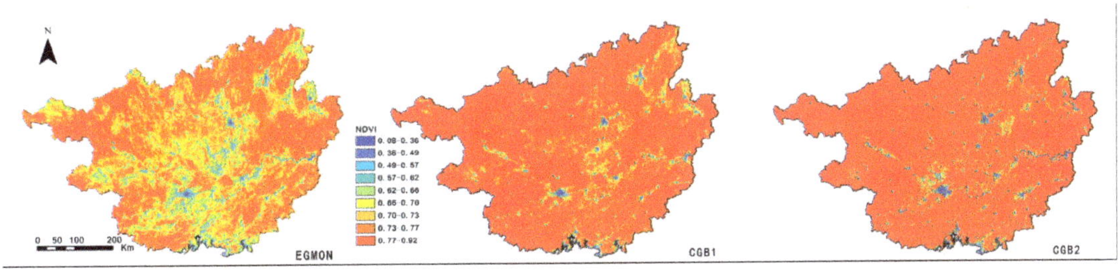


**Figure 5.** NDVI maps of EGMON, CGB1, and CGB2. NDVI: normalized difference vegetation index.

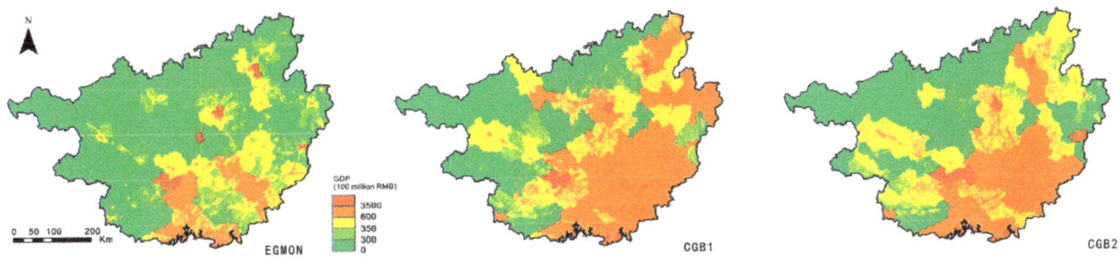


**Figure 6.** Gross domestic product (GDP) maps of EGMON, CGB1, and CGB2.

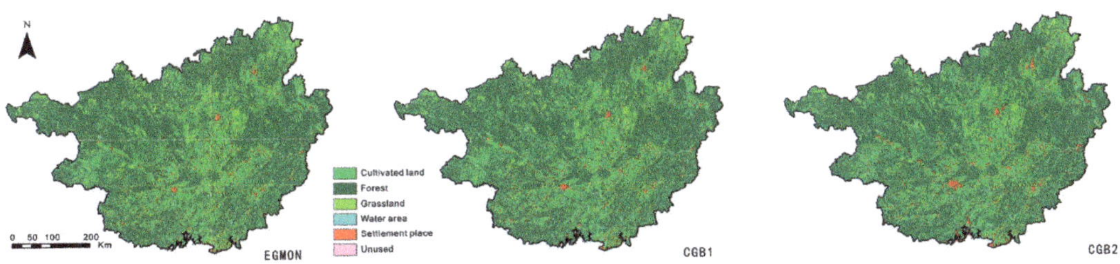


**Figure 7.** Land use maps of EGMON, CGB1, and CGB2.

*2.4. Random Forest Algorithm*

Random forest is a classic combined classifier algorithm proposed that was by Breiman [34]. A bagging algorithm was used to generate training sample subsets, and a classification regression tree was used as a meta-classifier, which was randomly selected to split the current node when building a single cart tree. This double random (random training set and random attribute) strategy results in a greater difference between meta-classifiers, which improves the classification performance [35].

Random forest is a combination of tree classifiers $\{h(x,\theta_k), k = 1, \ldots\ldots\}$, and its meta-classifier $h(x,\theta_k)$ is a complete growth and nonpruning classification regression tree. $x$ is the input vector, and $\{\theta_k\}$ is an independent and identically distributed random vector and determines the growth process of a single classification regression tree. For classification, the output of random forest is the result of simple majority voting, and the output is the simple average of the output of a single tree.

Random forest has many advantages; for example, the algorithm can incorporate both continuous and categorical attributes by using the cart algorithm as its meta-learning algorithm. Moreover, the decision tree is of great use in identifying differences and shows better classification performance owing to the combination of the bagging algorithm and randomly selected candidate feature splitting, which prevents overfitting and improves tolerance to noise.

Another prominent feature of random forest is the calculation of the importance of variables. First, the random forest algorithm adds disturbance by the random reordering of a variable of the training samples, and then it observes the change in the classification accuracy of all samples in the decision tree before and after disturbance to measure the variable importance.

Random forest analysis was completed using R3.4.2 [36]. The method used in this study is illustrated in Figure 8.

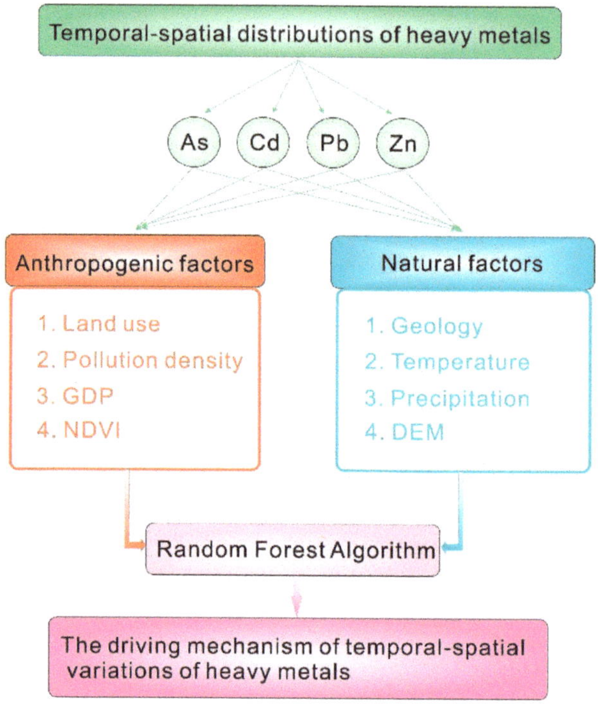


**Figure 8.** Flow chart showing the classification method used in this study.

## 3. Results

### *3.1. Descriptive Statistics*

Table 1 shows the statistical parameters of the As, Cd, Pb, and Zn concentrations in the EGMON, CGB1, and CGB2 projects. The maximum values of As, Cd, Pb, and Zn (1270.72 mg/kg, 30.61 mg/kg, 1385.64 mg/kg, and 3724.96 mg/kg respectively) for CGB1 were the highest, indicating that heavy metals were heavily enriched in the CGB1 project. In addition, the coefficients of variation of As, Cd, Pb, and Zn were 405.67, 370, 314.1, and 320.28, respectively, in the CGB1 project, which were also higher than those of the other two projects, suggesting that they had the strongest spatial variations. Moreover, As, Cd, and Pb showed the highest median values (12.58 mg/kg, 0.26 mg/kg, and 31.04 mg/kg, respectively) in the CGB1 project, and Zn showed the highest median value in the CGB2 project. The medians of As, Cd, and Zn were lowest in the EGMON project, and the median of Pb was lowest in CGB2.

**Table 1.** Descriptive statistics of heavy metals in the EGMON, CGB1, and CGB2 projects.

| | | As | Cd | Pb | Zn |
|---|---|---|---|---|---|
| EGMON | Mean | 15.77 | 0.17 | 38.35 | 85.98 |
| | CV | 193.49 | 70.45 | 72.28 | 79.51 |
| | Median | 7.80 | 0.13 | 34.50 | 72.00 |
| | Min | 2.10 | 0.05 | 14.00 | 17.50 |
| | Max | 144.00 | 0.47 | 137.00 | 342.00 |
| CGB1 | Mean | 59.71 | 1.57 | 83.06 | 218.96 |
| | CV | 405.67 | 370.00 | 314.10 | 320.28 |
| | Median | 12.58 | 0.26 | 31.04 | 79.42 |
| | Min | 3.12 | 0.10 | 16.10 | 37.96 |
| | Max | 1270.72 | 30.61 | 1385.64 | 3724.96 |
| CGB2 | Mean | 12.81 | 0.39 | 32.35 | 87.84 |
| | CV | 78.07 | 10 | 75.50 | 61.38 |
| | Median | 12.17 | 0.24 | 26.95 | 80.22 |
| | Min | 2.47 | 0.03 | 10.60 | 27.15 |
| | Max | 50.42 | 1.53 | 137.92 | 261.79 |

Units: mg/kg. CV: coefficient of variation.

### *3.2. Spatiotemporal Distributions of Heavy Metals*

The heavy metal geochemical maps of the three projects (Figure 9) were drawn using the inverse distance weighting method in ArcGIS 10.4 with the European Union soil heavy metal pollution standard values (indicated by the triangular symbol on the color scale) as the thresholds. Arsenic showed the strongest enrichment in the three projects. In the EGMON project, As-anomalous areas were mainly distributed in the east and west of the study area, and the proportion of alluvial soil samples exceeding the limit of 20 mg/kg of As was approximately 10%. As-anomalous areas in the CGB1 project were distributed in the northwest and southeast of the study area, and the proportion exceeding this limit was approximately 23%. In addition, the As-anomalous areas in the CGB2 project were mainly distributed in the northwest and northeast of the study area, showing an exceeding proportion of 11%, and the anomalous intensity and range in the CGB1 project were both weaker than those in the CGB1 project.

In the EGMON project, the Cd content at all sampling sites was below the Cd EU heavy metal pollution threshold of 1000 ug/kg. In the CGB1 project, the Cd-anomalous areas were mainly distributed in the northwest of the study area, with an exceeding proportion of 17%, while the Cd-anomalous intensity and range decreased significantly, showing an exceeding proportion of 7% in the CGB2 project.

The anomalous areas of Pb and Zn in the EGMON project were mainly distributed in the northwest and southeast of the study area, and the proportions of alluvial soil samples exceeding the limits of 70 mg/kg Pb and 160 mg/kg Zn were both approximately 5%. The exceeding proportions of Pb and Zn in the CGB1 project were approximately 17% and 13%, respectively. Compared with the previous two projects, the anomalous intensity and range of Pb in the CGB2 project decreased significantly, with an exceeding proportion of 3.5%, and the anomalous intensity and range of Zn in the CGB2 project were significantly lower than those in the CGB1 project but higher than those in the EGMON project, with an exceeding proportion of 7%.

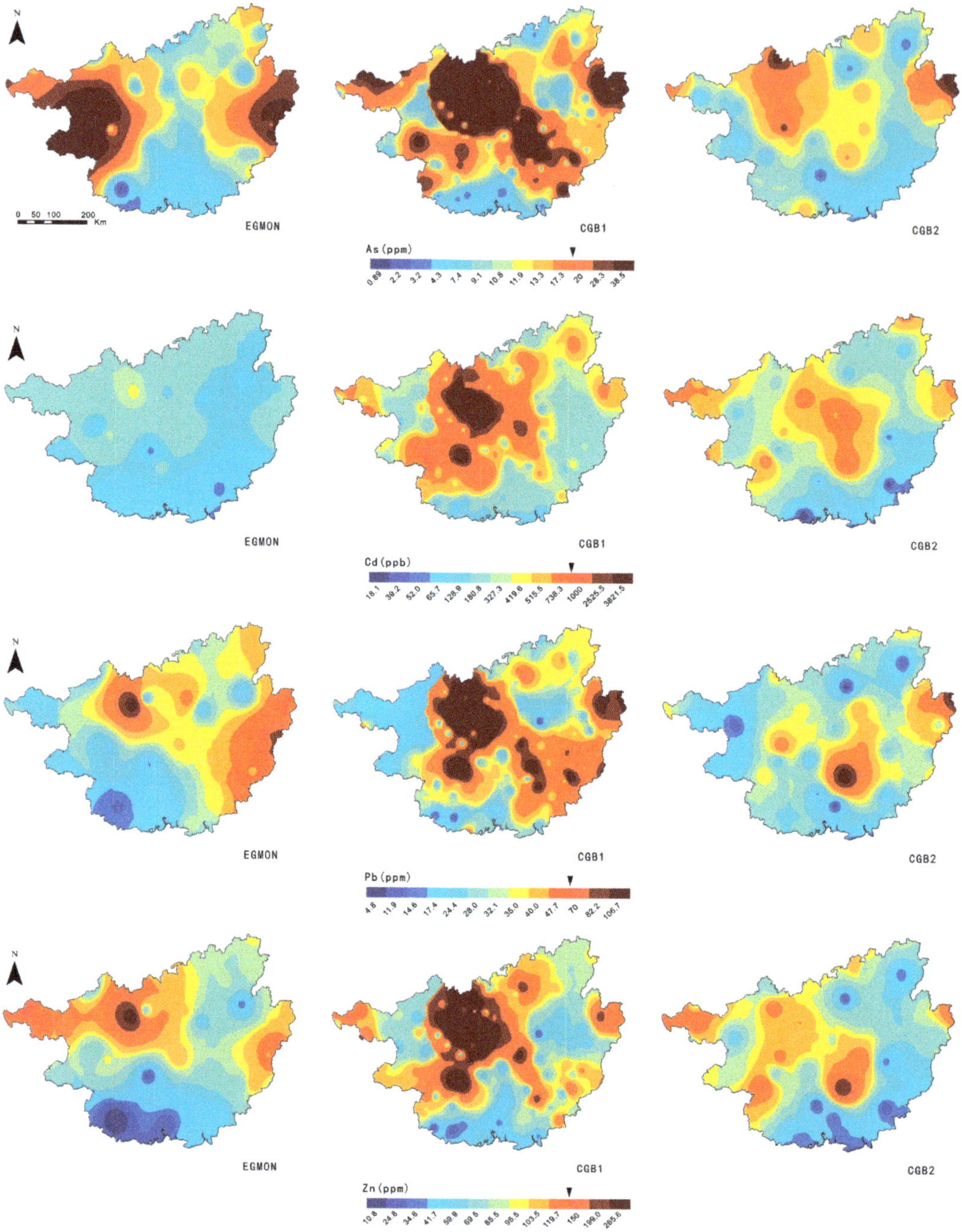


**Figure 9.** Geochemical maps of As, Cd, Pb, and Zn concentrations in Guangxi. The triangle indicates the threshold of EU heavy metal pollution standard.

## 4. Discussion

Land use affects soil quality and eco-environmental function, which may be related to soil chemical availability and environmental migration conditions as well as fertilization and irrigation under different land use conditions [37,38]. Gross Domestic Product likely reflects the environmental problems caused by rapid economic development, including industrial and commercial development [39]. The density of the pollution sources around the sampling sites directly reflects the influence of industrial emissions on the heavy metal content of the soils. Pollution can contaminate the downstream soils through atmospheric sedimentation and surface runoff [40,41]. The NDVI is a positive anthropogenic factor that indicates an improved ecological environment and reduced rocky desertification in karst areas [42]. Natural heavy metals in the soil originate from weathering, erosion, and the transport of parent materials. Geological lithology likely reflects the type of parent material. Temperature and precipitation are climatic factors that determine the weathering and denudation rate of parent materials, and lastly, the DEM can determine the path and difficulty of element migration.

The variable importance of the influencing factors, calculated using the random forest algorithm, is illustrated in Figure 10. The spatial variations in As in the EGMON project were mainly affected by precipitation, followed by the spatial distribution density of pollution sources and GDP. The As-anomalous areas geographically coincided with poor-precipitation areas. The greater the rainfall, the stronger the weathering and leaching, while the batholith ions in the soil are more easily lost; thus, the soil is acidic. Fe–Mn concretions are often associated with rich heavy metals because of their poor crystallinity, large surface area [43], high surface negative charge, and the isomorphic substitution of manganese oxides. Fe–Mn concretions in the soil decompose through the acidification process, which leads to the migration of heavy metals from the solid state to the ionic state, resulting in a relatively low heavy metal content in the soil. Natural factors were the key factors regulating the spatial variation of As, with a proportion of more than 60% in the EGMON project. As spatial variation was controlled by GDP and the spatial distribution density of pollution sources in the CGB1 project, the proportion of anthropogenic factor importance was approximately 58%. China experienced rapid economic development during this period, and extensive economic growth has brought many problems, such as the wanton discharge of pollutants and resulting environmental pollution, which were the key factors controlling the spatial distribution of As in the CGB1 project. Arsenic is mainly derived from smelting, pigments, glass, and paper manufacturing in industry, and in agriculture, arsenide is mainly used in pesticides, algicides, and preservatives [44]. The As-anomalous areas were decreased in CGB2 compared with CGB1, and the proportion of anthropogenic factor importance was approximately 55%. The results of the random forest analysis showed that the pollution source was still the key factor affecting the As spatial distribution; however, the NDVI took second place and should be given more focus for the positive role that it plays. The NDVI of the study area increased significantly in the CGB2 project compared with the previous two projects. Vegetation likely regulates surface runoff and conserves soil and water, thereby improving soil quality and the ecological environment. Therefore, it can be concluded that a higher NDVI is associated with a lower degree of rocky desertification; thus, the soil erosion and pollutant contents decreased correspondingly, reflecting the positive impact of human activities on the spatial distributions of heavy metals [45,46].

The key factors regulating the Cd spatial distribution in the EGMON project were natural factors, including DEM, temperature, and geological lithologies, the proportion of which was 61%. The Cd-anomalous areas geographically coincided with low-lying areas; Cd is an active element that migrates more easily than other heavy metals, and low-lying areas can gather materials eroding from the surrounding high-relief areas, resulting in the accumulation of Cd. GDP was the most important factor regulating the spatial distribution of Cd in the CGB1 project, with a proportion of anthropogenic factors of approximately 53%. Cd is a by-product of Zn smelting, which is mainly used in batteries, dyes, and plastic stabilizers, and is more easily absorbed by crops than other heavy metals [47]. The anomalous intensity and range of Cd in CGB2 decreased significantly, with precipitation being the main factor affecting its spatial distribution.

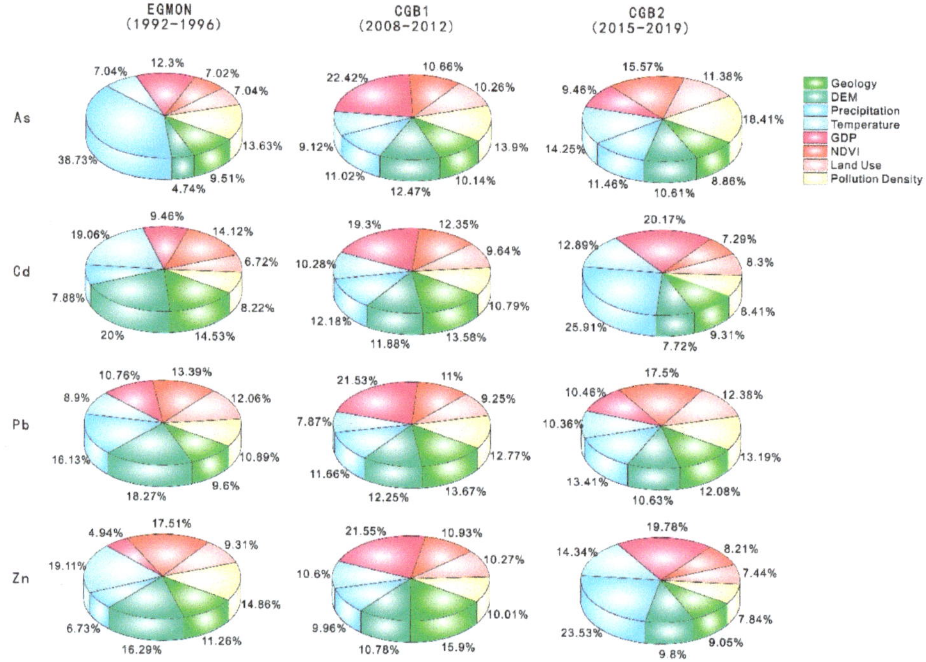


**Figure 10.** Influencing factors of spatiotemporal distributions of heavy metals.

The spatial variation in Pb in the EGMON project was mainly controlled by the DEM and precipitation, and the proportion of natural factors was approximately 53%. High wet deposition was accompanied by high soil Pb content in Guangxi in the EGMON project, which was different from the other elements. Leaded gasoline was used in China during the EGMON project period; thus, Pb from the exhaust emissions of motor vehicles using leaded gasoline would be adsorbed onto particles and enter the soil through wet deposition [48]. Gross Domestic Product is an important factor controlling the spatial distribution of Pb in the CGB1 project. During this period, leaded gasoline was completely banned in China; thus, lead in soils mainly originated from mining, smelting, leaded coatings, foundry, and other industrial production activities [49]. In the CGB2 project, NDVI was the controlling factor affecting spatial variations in Pb, suggesting a positive role for human activities. The implementation of vegetation restoration and the control of rocky desertification decreased the contributions of upstream rocks; thus, the heavy metal content in the alluvial soils decreased accordingly.

The spatial variation in Zn in the EGMON project was mainly affected by temperature, and the proportion of natural factor importance was approximately 53%. There was a negative correlation between Zn content and temperature. Zn is more active at high temperatures and easily migrates from the solid state to the ionic state, resulting in relatively low Zn content in soils. In the CGB1 project, the GDP was still a key factor, and the proportion of anthropogenic factor importance was greater than 53%. Zn pollution sources include zinc mining, smelting and processing, machinery manufacturing, zinc plating, instrumentation, synthesis, and papermaking. Tire wear and coal combustion also produce zinc and zinc compounds. Precipitation was the most important factor affecting the spatial variation in Zn in the CGB2 project, whereas GDP still played an important role. It was found that the Zn-anomalous areas changed compared with the CGB1 period, which might have been due to the development of the local economy.

In conclusion, the surface soil environmental quality in Guangxi was the best during the EGMON project (1992–1996) period. The disturbance of human activities was relatively small, and the spatial distribution of heavy metals was mainly driven by natural factors. In the CGB1 project period (2008–2012), surface soils showed obvious heavy metal pollution due to rapid economic growth and poor attention to environmental protection. In the CGB2 project (2015–2019), economic growth slowed down, environmental protection was strengthened, the degree of rocky desertification was reduced, and the phenomenon of soil erosion was greatly improved; thus, the anomalous intensity and range of heavy metals decreased compared with that in the CGB1 project.

The factors influencing the spatial distributions of Cd and Zn were similar (Figure 9). Cd is a dispersed element that is closely associated with middle- to low-temperature Pb–Zn deposits and occurs in sphalerite, wurtzite, and other minerals as isomorphisms. Both Cd and Zn are sulfophilic elements with similar ionic radii; therefore, their geochemical behavior is consistent. It is generally accepted that, at the local scale, soil-forming parent materials are among the most important factors affecting heavy metal distribution in soils [50]. However, this study found that wet deposition associated with the DEM is among the most important factors responsible for the spatial distribution of heavy metals in karst areas at a regional scale. The intense tropical rainfall associated with the special geochemical properties of carbonates may be the key factor controlling mass migration and, thus, heavy metal accumulation in karst areas.

Spatial variations in heavy metal concentrations are caused by many factors, including numerical variables (such as GDP and NDVI) and categorical variables (such as geology background and land use). In the past, owing to the limitations of methods, most studies have only focused on numerical variables of the spatial variations of heavy metals, lacking a comprehensive study of both numerical and categorical variables. In this study, numerical and categorical variables were evaluated for the first time using a random forest algorithm. This research promotes an understanding of anthropogenic and natural influences on the spatiotemporal distribution of heavy metals and is of great significance for environmental monitoring and governance. It is of great significance to study the distribution of heavy metals on large spatial and temporal scales, as well as the potential influencing factors. It is necessary to monitor the spatial and temporal changes in heavy metals on a regional scale over long time periods and clarify the impact of human disturbance on natural environmental change, which is also one of the original intentions of long-term geochemical projects in China.

## 5. Conclusions

In conclusion, the surface alluvial soils showed the strongest heavy metal accumulation in the CGB1 project, followed by the CGB2 and EGMON projects. Arsenic showed the strongest enrichment among the three projects. Natural factors were among the most important factors affecting the spatial distribution of heavy metals in the EGMON project, particularly precipitation for As, DEM for Cd and Pb, and temperature for Zn. Gross Domestic Product was a key factor regulating the spatial distribution of all heavy metals in CGB1. The NDVI and precipitation were important factors controlling heavy metal variations in CGB2. As a positive anthropogenic factor, the NDVI improved the degree of rocky desertification, reduced the heavy metal contents of As and Pb, and promoted the decomposition of oxides and hydroxides and thus the migration of Cd and Zn.

**Author Contributions:** Conceptualization, M.T.; methodology, M.T.; software, M.T.; formal analysis, M.T.; investigation, X.W., J.Z. and Q.C.; resources, D.L.; data curation, M.T., X.W., J.Z., Q.C., D.L., Y.Q., C.W., Y.Z., Q.W. and F.L.; writing—original draft preparation, M.T.; writing—review and editing, Q.W.; visualization, Y.Q., C.W., Y.Z., Q.W. and F.L.; supervision, J.Z. All authors have read and agreed to the published version of the manuscript.

**Funding:** This work was supported by the National Natural Science Foundation (41907290), Global Geochemical Baselines Project (DD20230013, DD20190450, DD 20160116, Sinoprobe-04), Surplus Funds of Institute of Geophysical and Geochemical Exploration (JY201909) and National Nonprofit Institute Research Grant of IGGE (AS2022P03).

**Acknowledgments:** Many thanks are given to all participants in the China Geochemical Baselines projects. Comments by anonymous reviewers considerably improved the paper.

**Conflicts of Interest:** The authors declare no conflict of interest.

## References

1. Arfaeinia, H.; Dobaradaran, S.; Moradi, M.; Pasalari, H.; Mehrizi, E.A.; Taghizadeh, F.; Esmaili, A.; Ansarizadeh, M. The effect of land use configurations on concentration, spatial distribution, and ecological risk of heavy metals in coastal sediments of northern part along the Persian Gulf. *Sci. Total Environ.* **2019**, *653*, 783–791. [CrossRef] [PubMed]
2. Liu, H.; Zhang, Y.; Yang, J.; Wang, H.; Li, Y.; Shi, Y.; Li, D.; Holm, P.; Ou, Q.; Hu, W. Quantitative source apportionment, risk assessment and distribution of heavy metals in agricultural soils from southern Shandong Peninsula of China. *Sci. Total Environ.* **2021**, *767*, 144879. [CrossRef] [PubMed]
3. Wu, W.; Wu, P.; Yang, F.; Sun, D.; Zhang, D.; Zhou, Y. Assessment of heavy metal pollution and human health risks in urban soils around an electronics manufacturing facility. *Sci. Total Environ.* **2018**, *630*, 53. [CrossRef] [PubMed]
4. Yang, Q.; Li, Z.; Lu, X.; Duan, Q.; Lei, H.; Bi, J. A review of soil heavy metal pollution from industrial and agricultural regions in China: Pollution and risk assessment. *Sci. Total Environ.* **2018**, *642*, 690–700. [CrossRef] [PubMed]
5. Zhang, C.; Yang, Y. Modeling the spatial variations in anthropogenic factors of soil heavy metal accumulation by geographically weighted logistic regression. *Sci. Total Environ.* **2020**, *717*, 137096. [CrossRef] [PubMed]
6. Wen, Y.; Li, W.; Yang, Z.; Zhuo, X.; Guan, D.; Song, Y.; Guo, C.; Ji, J. Evaluation of various approaches to predict cadmium bioavailability to rice grown in soils with high geochemical background in the karst region, southwestern China. *Environ. Pollut.* **2020**, *258*, 113645. [CrossRef] [PubMed]
7. Sullivan, P.L.; Macpherson, G.L.; Martin, J.B.; Price, R.M. Evolution of carbonate and karst critical zones. *Chem. Geol.* **2019**, *527*, 119–223. [CrossRef]
8. Li, Z.; Ma, Z.; Kuijp, T.J.; Yuan, Z.; Huang, L. A review of soil heavy metal pollution from mines in China: Pollution and health risk assessment. *Sci. Total Environ.* **2014**, *468–469*, 843–853. [CrossRef]
9. Li, P.; Zhi, Y.; Shi, J.; Zeng, L.; Wu, L. County-scale temporal-spatial distribution and variability tendency of heavy metals in arable soils influenced by policy adjustment during the last decade: A case study of Changxing, China. *Environ. Sci. Pollut. Res.* **2015**, *22*, 17937–17947. [CrossRef]
10. Nickel, S.; Hertel, A.; Pesch, R.; Schroder, W.; Steinnes, E.; Uggerud, H. Modelling and mapping spatio-temporal trends of heavy metal accumulation in moss and natural surface soil monitored 1990-2010 throughout Norway by multivariate generalized linear models and geostatistics. *Atmos. Environ.* **2014**, *99*, 85–93. [CrossRef]
11. Rodríguez Martín, J.A.; De Arana, C.; Ramos-Miras, J.J.; Gil, C.; Boluda, R. Impact of 70 years urban growth associated with heavy metal pollution. *Environ. Pollut.* **2015**, *196*, 156–163. [CrossRef] [PubMed]
12. Shukla, L.; Jain, N. A review on soil heavy metals contamination: Effects, sources and remedies. *Appl. Ecol. Environ. Sci.* **2022**, *10*, 15–18. [CrossRef]
13. Tian, M.; Wang, X.; Liu, F.; Hu, Q.; Qiao, Y.; Wang, Q. Spatial-temporal variability and influence factors of Cd in soils of Guangxi, China. *PLoS ONE* **2023**, *18*, e0279980. [CrossRef] [PubMed]
14. Xie, X.J.; Cheng, H.X. The suitability of floodplain sediment as a global sampling medium: Evidence from China. *J. Geochem. Explor.* **1997**, *58*, 51–62. [CrossRef]
15. Xie, X.J.; Ma, X.Z.; Ren, T.X. Geochemical mapping in China. *J. Geochem. Explor.* **1997**, *60*, 99–113.
16. Wang, X. Global Geochemical Baselines: Understanding the past and predicting the future. *Earth Sci. Front.* **2012**, *19*, 7–18. (In Chinese with English Abstract)
17. Wang, X.Q. China geochemical baselines: Sampling methodology. *J. Geochem. Explor.* **2015**, *148*, 25–39. [CrossRef]
18. Wang, X.; Zhou, J.; Xu, S.; Chi, Q.; Nie, L.; Zhang, B.; Yao, W.; Wang, W.; Liu, H.; Liu, D.; et al. China soil geochemical baselines networks: Data characteristics. *Geol. China* **2016**, *43*, 1469–1480. (In Chinese with English Abstract)
19. Tian, M.; Wang, X.; Qiao, Y.; Liu, D.; Chi, Q.; Liu, H.; Wang, W.; Zhang, B. Temporal Variations of Sediment Provenance in a Karst Watershed, China. *Appl. Sci.* **2023**, *13*, 771. [CrossRef]
20. Karim, Z.; Qureshi, B.A.; Mumtaz, M.; Salmon, Q. Heavy metal content in urban soils as an indicator of anthropogenic and natural influences on landscape of Karachid A multivariate spatio-temporal analysis. *Ecol. Indic.* **2014**, *42*, 20–31. [CrossRef]
21. Wong, S.C.; Li, X.; Zhang, G.; Qi, S.; Min, Y. Heavy metals in agricultural soils of the pearl River Delta, south China. *Environ. Pollut.* **2002**, *119*, 33–44. [CrossRef] [PubMed]
22. Sparks, D.L. *Environmental Soil Chemistry*, 2nd ed.; Academic Press: San Diego, CA, USA, 2003.
23. Cheng, W.; Yang, Z.F.; Cong, Z.; Ji, J.F. Temporal-spatial variation and source apportionment of soil heavy metals in the representative river-alluviation depositional system. *Environ. Pollut.* **2016**, *216*, 18–26.

24. Liang, S.Y.; Cui, J.L.; Bi, X.Y.; Luo, X.; Li, X. Deciphering source contributions of trace metal contamination in urban soil, road dust, and foliar dust of Guangzhou, southern China. *Sci. Total Environ.* **2019**, *695*, 133596.1–133596.12. [CrossRef] [PubMed]
25. Ren, S.; Song, C.; Ye, S.; Cheng, C.; Gao, P. The spatiotemporal variation in heavy metals in China's farmland soil over the past 20 years: A meta-analysis. *Sci. Total Environ.* **2022**, *806*, 150322. [CrossRef] [PubMed]
26. Breiman, L. Random forests. *Mach. Learn.* **2001**, *45*, 5–32. [CrossRef]
27. Cracknell, M.J.; Reading, A.M.; Mcneill, A.W. Mapping geology and volcanic-hosted massive sulfide alteration in the Hellyer-Mt Charter region, Tasmania, using Random Forests™ and Self-organising maps. *J. Geol. Soc. Aust.* **2014**, *61*, 287–304. [CrossRef]
28. Wiesmeier, M.; Steffens, M.; Kölbl, A.; Knabner, I. Degradation and small-scale spatial homogenization of topsoils in intensively grazed steppes of Northern China. *Soil Tillage Res.* **2009**, *104*, 299–310. [CrossRef]
29. Carranza, E.J.M.; Laborte, A.G. Random forest predictive modeling of mineral prospectivity with small number of prospects and data with missing values in Abra (Philippines). *Comput. Geosci.* **2015**, *74*, 60–70. [CrossRef]
30. Darnley, A.G.; BjorklundJ, A.; Boviken, B. *A Global Geochemical Database for Environmental and Resource Management: Final Report of IGCP Project 259, Earth Sciences, 19*; UNESCO Publishing: Paris, France, 1995; 122p.
31. Lark, R.M.; Bellamy, P.H.; Rawlins, B.G. Spatio-temporal variability of some metal concentrations in the soil of eastern England, and implications for soil monitoring. *Geoderma* **2006**, *133*, 363–379. [CrossRef]
32. Zhang, Q.; Bai, J.; Wang, Y. Analytical scheme and quality monitoring system for China Geochemical Baselines. *Earth Sci. Front.* **2012**, *19*, 33–42. (In Chinese with English Abstract)
33. Xie, X.J.; Yan, M.C.; Li, L.Z.; Shen, H. Usable values for Chinese standard reference samples of stream sediments, soils, and rocks: GSD 9–12, GSS 1–8 and GSR 1–6. *Geostand. Newsl.* **1985**, *9*, 277–280.
34. Breiman, L. Bagging predictors. *Mach. Learn.* **1996**, *24*, 123–140. [CrossRef]
35. Tian, M.; Wang, X.; Nie, L.; Liu, H.; Wang, W.; Yan, T. Spatial distributions and the identification of ore-related anomalies of Cu across the boundary area of China and Mongolia. *J. Geochem. Explor.* **2019**, *197*, 37–47. [CrossRef]
36. Team, R.R.D.C. A language and environment for statistical computing. *Computing* **2007**, *1*, 12–21.
37. Yang, T.; Han, G. Characteristics of heavy metals in soils under different land use in a typical karst area, southwest China. *Acta Geochim.* **2017**, *36*, 515–518.
38. Shu, X.; Li, Y.; Li, F.; Feng, J.; Shen, J.; Shi, Z. Impacts of land use and landscape patterns on heavy metal accumulation in soil. *Environ. Sci.* **2019**, *40*, 2471–2482. (In Chinese with English Abstract)
39. Liu, L.L.; Liu, Q.Y.; Ma, J.; Wu, H.; Qu, Y.; Gong, Y.; Yang, S.; An, Y.; Zhou, Y. Heavy metal(loid)s in the topsoil of urban parks in Beijing, China: Concentrations, potential sources, and risk assessment. *Environ. Pollut.* **2020**, *260*, 114083. [CrossRef]
40. Gao, Y.F.; Liu, H.L.; Liu, G.X. The spatial distribution and accumulation characteristics of heavy metals in steppe soils around three mining areas in Xilinhot in Inner Mongolia, China. *Environ. Sci. Pollut. Res.* **2017**, *24*, 16–25. (In Chinese with English Abstract) [CrossRef]
41. Solgi, E.; Parmah, J. Analysis and assessment of nickel and chromium pollution in soils around Baghejar Chromite Mine of Sabzevar Ophiolite Belt, Northeastern Iran. *Trans. Nonferrous Met. Soc. China* **2015**, *25*, 2380–2387. [CrossRef]
42. Qiao, P.W.; Yang, S.C.; Lei, M. Quantitative analysis of the factors influencing spatial distribution of soil heavy metals based on geographical detector. *Sci. Total Environ.* **2019**, *664*, 392–413. [CrossRef]
43. Wang, P.; Wang, S.L.; Liu, S.Q.; Li, Y.; He, M.; Lin, X. Occurrence, speciation, source and geochemical cycle of Arseni. *Environ. Sci. Technol.* **2010**, *7*, 96–103. (In Chinese with English Abstract)
44. Bo, Y.; Luo, L.Q. Geochemical characteristics and research direction of Arsenic. *Rock Miner. Anal.* **2009**, *28*, 569–575, (In Chinese with English Abstract).
45. Li, Z.; Xu, X.; Zhang, Y.; Wang, K. Fingerprinting sediment sources in a typical karst catchment of southwest China. *Int. Soil Water Conserv. Res.* **2020**, *8*, 277–285. [CrossRef]
46. Guan, Z.; Tang, X.Y.; Yang, J.E.; Ok, Y.S.; Xu, Z.; Nishimura, T.; Reid, B.J. A review of source tracking techniques for fine sediment within a catchment. *Environ. Geochem. Health* **2017**, *39*, 1–23. [CrossRef]
47. Zhang, X.; Yan, Y.; Wadood, S.A.; Sun, Q.; Guo, B. Source apportionment of cadmium pollution in agricultural soil based on cadmium isotope ratio analysis. *Appl. Geochem.* **2020**, *123*, 104776. [CrossRef]
48. Zhu, H.Y.; Jin, H.; Liu, R. Lead and its variation in wet deposition of Qingdao. *Environ. Monit. China* **2001**, *17*, 45, (In Chinese with English Abstract).
49. He, B.; Zhao, X.; Li, P.; Liang, J.; Fan, Q.; Ma, X.; Zheng, J.; Qiu, J. Lead isotopic fingerprinting as a tracer to identify the pollution sources of heavy metals in the southeastern zone of Baiyin, China. *Sci. Total Environ.* **2019**, *660*, 348–357. [CrossRef] [PubMed]
50. Su, Z.; Wen, X.F.; Wu, P.; Liu, H.; Liu, Y.; Pan, Q.; Wei, X.; Wu, S. Excessive Degrees and Migration Characteristics of Heavy Metals in Typical Weathering Profiles in Karst Areas. *Earth Environ.* **2019**, *47*, 53–59, (In Chinese with English Abstract).

minerals

MDPI

Article

# Development of the Recycling of Alloyed Metallurgical Waste: Features of Phase and Structural Transformations

**Aleksey Smirnov [1], Artem Petryshchev [2], Tamara Bilko [3], Andrey Andreev [4], Anastasiia Semenko [1,*] and Yuliia Skorobagatko [1]**

1 Physico-Technological Institute of Metals and Alloys, The National Academy of Science of Ukraine (PTIMA NASU), 03142 Kyiv, Ukraine; stalevoz@i.ua (A.S.); mgd@ptima.kiev.ua (Y.S.)

2 Department of Labor and Environment Protection, National University "Zaporizhzhia Polytechnic", 69063 Zaporizhzhia, Ukraine; kafedrales@ukr.net

3 Department of Labor Protection and Biotechnical Systems, National University of Life and Environmental Sciences of Ukraine, 03041 Kyiv, Ukraine; bilko@nubip.edu.ua

4 Department of General and Applied Physics, Zaporizhzhia National University, 69600 Zaporizhzhia, Ukraine; andreevandrijn@gmail.com

* Correspondence: semenko.au@gmail.com

**Abstract:** The features of structural and phase transformations during the processing of alloyed metallurgical wastes using reduction smelting are determined herein. This is necessary in order to determine the technological parameters of the melting process that ensure the reduction in the losses of alloying components. The use of X-ray phase analysis in combination with the methods of raster electron microscopy and X-ray microanalysis ensured the identification of the microstructure features and the chemical composition of individual phases and inclusions in the metal. Our study identified new technological aspects of high-alloyed technogenic waste processing using reduction smelting. The obtained parameters of the resource-saving alloying compound provide the possibility to replace parts of the standard ferroalloys in steelmaking processes.

**Keywords:** oxide metallurgical waste; alloy steel scale; reduction smelting; phase transformations; microstructure

## 1. Introduction

The gradual depletion of the raw materials fields dictates the rising trend of the global market prices for refractory metal alloys [1]. One of the alternative sources of alloying materials is processing and returning to the production of alloyed technogenic wastes, large volumes of which are not used effectively in practice [2]. Wastes of alloyed, heat-resistant, corrosion-resistant, and other steel and alloy grades, the operation of which can be accompanied by the effects of aggressive atmospheres, temperatures, and mechanical factors, contain expensive chemical elements. These elements include Ni, Cr, W, Mo, and many others. A specific feature of wastes is the presence of alloying elements in the form of oxides and complex compounds. This requires consideration of the complexity of chemical–physical interactions of elements when developing the technological conditions of processing [3]. The major portion of the alloyed wastes contains oxide and fine-dispersed compounds (scale, power grinding chips, and grinding dust), the effective processing of which is complicated. The above characteristics present challenges in terms of implementing the competitive ability due to problems with production processes and the high price of the finished products, respectively [4].

Therefore, the problem with decreasing losses of expensive chemical elements in wastes processing resulting from the production of high-alloy steels and compounds still needs to be urgently addressed. To accomplish this, transformations during the processing of metallurgical oxide industrial wastes using reduction smelting should be studied.

In accordance with the paper cited in [5], 0.043 t of scale and slag accounts for 1 ton of produced rolled steel. Steel losses, when it is processed in scarfing machines, are 5%–15%. The particle size for stainless corrosion-resistant steels with a nickel content of 19 wt% is 240–450 μm. Based on the data from the work cited in [6], in the best case scenario, 12% of the manufactured product volume is lost with the scale when small commercial billets are made of high-alloy steel. The difficulties lie in the fact that pre-treatment of scale and other fine-dispersed wastes is required in order to increase the level of alloy element recovery by molten metal. Moreover, there are additional difficulties associated with "oily" (contaminated) scale [7]. This means that before adding to the charge, harmful admixtures must be treated off. Based on the data in [6], when finely-dispersed high-alloy steel waste is added to a liquid bath without pre-treatment, the loss of refractory elements reaches 40%.

Iron scale is composed of $Fe_3O_4$, $Fe_2O_3$, and FeO, as specified in [8], but [6] also identified $WO_2$ and $W_2C{\cdot}Mo_2C$ in the high-speed steel scale. The reason for this may be an increased level of steel alloying. Similar results to those for the iron carbide presence were shown in [9], when chromium-nickel steel oxide waste was reduced by the carbon. The authors of the study in [8] investigated the iron scale reduction by carbon within the range of 1023–1323 K, with the detection of $Fe_3C$ and carbon along with the iron phase in the resultant products. Residual iron oxides appeared only after heat treatment at a temperature of 1023 K. In contrast, as indicated in [10], the additional presence of tungsten seems to have had an effect on the nature of carbide formations, which affected the microstructure's features. Some particles, probably carbide ones, with high levels of carbon, tungsten, and other elements, were found. Among the shortcomings, insufficient research into the reduction products' physical and chemical properties involving refractory elements—tungsten and niobium—can be noted. There is still a need to identify the most preferable conditions for reducing alloyed technogenic raw materials in the Fe-Ni-Cr-Mo-W-O-C system.

The reduction of oxides $FeO$-$Cr_2O_3$ at different C:Fe ratios and temperatures varying from 1373 K to 1523 K was studied by the authors of [11]. It was found that the increase in the C:Fe ratio from 0.8 to 1.4 caused the increase in the chromium extraction rate from 9.6% to 74.3%, respectively. The temperature rise to 1523 K caused the formation of a larger quantity of carbides, the residues of which could not be avoided in the products reduced by carbon [12]. If the C:Fe ratio was less than 0.8, there was a significant decrease in the chromium extraction rate and the carbide formation. The resulting chromium carbides dissolved in the iron phase [11]. The formation of metallic chromium and chromium carbides was also demonstrated by the authors of [13] during the reduction of chromium-containing ore raw materials. After heat treatment at a temperature of 1273 K and reaching the O:C ratio of 1.05–1.15 in the charge, the carbothermal reduction was accompanied by the occurrence of $Cr_3C_2$ and $Cr_7C_3$. Thus, to increase the chromium extraction rate during the reduction, the charge must contain certain excessive carbon relative to oxygen. The study cited in [14] specifically researches $Cr_2O_3$ reduction with carbon within a range of 1273–1773 K. Parallel reduction and formation of $Cr_3C_2$, $Cr_7C_3$, and $Cr_{23}C_6$ carbides were found, and the possibility of obtaining iron–chromium ligatures with the limited carbon content was fixed. There is a possibility of obtaining products of carbon thermal reduction with relatively low residual carbon, which indicates the expanded scope of application. However, it is impossible to see the regularities of changes in the microstructure and composition of the phase formations in the obtained materials during the reduction of the chromium and iron oxides complex within the alloyed technogenic raw materials.

The study cited in [15] outlines the research into reduction reactions in the Fe–Ni–O system at a temperature of 1373 K. When the treatment temperature rose, the reduced products appeared in the following sequence: $Fe_3O_4 + NiO \rightarrow Fe_3O_4 + Ni \rightarrow FeO + Fe_xNi_y \rightarrow Fe_xNi_y + Fe\ (Fe_nC_m)$. That is, a relatively considerable tendency for nickel oxide reduction against iron oxides was observed. The formation processes of metallic nickel and iron were seen in separate stages. The possible presence of iron carbides and the iron-containing phase $Fe_xNi_y$, which can appear during the reduction of oxide alloyed waste, was indicated.

The disadvantage lies in the absence of data regarding the reduction of complex oxide compounds which can be present in technogenic raw materials. This may lead to potential differences during the reduction reactions. The unresolved parts of the problem lie in the expansion of our understanding of the elements' presence in reduced products, determined using scanning electron microscopy and X-ray microanalysis.

The authors of [16] studied the reduction in the Mo–O–C and Ca–Mo–O–C systems. The formation of molybdenum dioxide as an intermediate product was found; the latter was further transformed into molybdenum and carbides. The conditional division of the process into the primary interaction of $MoO_3$ with carbon occurred with the carbon gasification reaction, as did the further reduction of $MoO_2$ with the involvement of carbon monoxide. The authors of [17] made calculations using thermodynamics and determined the equilibrium of $WO_3$ reduction reactions with the involvement of carbon and carbon monoxide at a temperature of 1500–6000 K. The obtained characteristics of the reactions showed that the reduction of $WO_3$ most likely occurred to produce tungsten and then the carbides $W_2C$ and WC. As was indicated in [18], there are stages of $WO_3$ transformation through $WO_{2,72}$ and $WO_2$ to W. The tungsten reduction takes place at a temperature of 1223 K. The tendency to form carbides, along with the reduction of oxide tungsten-containing compounds, was confirmed by the studies contained in [19]. The disadvantage is that the form of the molybdenum- and tungsten-containing compounds present in the oxide wastes of high-alloy steels can be more complicated and differ from specific pure oxides.

Based on the above, the purpose of this work was formulated. It lies in the identification of features of phase and structural transformations in the processing of high-alloy wastes from metallurgical production using reduction smelting to obtain an alloy containing refractory elements such as Ni, Cr, Mo, and W. This is necessary to determine the parameters that reduce the losses of alloying elements by means of sublimation in the processing of oxide-alloying raw materials and use of the received alloying additive.

## 2. Materials and Methods

In this study, a mixture of chromium–nickel-containing steels, type 18–10 (grades 08H18N10, 04H18N10 and others, respectively, GOST 5632-2014) scale, and a grinding chip of nickel-based heat-resistant alloys (grades EI868 and EK171, respectively, GOST 5632-2014) formed under metallurgical production conditions, was used as a feedstock. The addition of metal chips ensured the intensification of heat exchange at the initial stages of charge heating and an additional alloying improvement. The reducing agent was carbon in the form of ultradispersed dust, which was the waste from the carbon–graphite production, the addition of which ensured the O:C ratio in the charge at the level of 1.56. The samples for the research were melted in an indirect heating furnace with coal lining in alundum crucibles. The melting temperature was 1873–1913 K. After melting, the alundum crucibles with the alloy were taken out of the furnace and cooled at ambient temperature.

To make a comparative assessment of the probability of reduction reactions in the Ni-Cr-Mo-W-O-C system at a temperature of 300–2000 K, the Gibbs energy change ($\Delta G$) was calculated using the given thermodynamic values of the changes in enthalpy ($\Delta H$), entropy ($\Delta S$), and heat capacity ($\Delta Cp$) [20–25]. In the calculations, we used the technique described in [21], along with the following equation:

$$\Delta G_T^0 = \Delta H_{T_1}^0 + T\Delta S_{T_1}^0 + \int_{T_1}^{T} \Delta C_p dT - T\int_{T_1}^{T} \frac{\Delta C_p}{T} dT \quad (1)$$

The effect of the heat capacity changes and polymorphic transformation progress in the initial components and reaction products at different temperatures was also taken into account.

X-ray phase analysis of the samples was performed with the help of a diffractometer, using monochromatic radiation Cu K$\alpha$ ($\lambda$ = 1.54051 Å). Measurements were made at the tube voltage of U = 40 kV and anode current of I = 20 mA. Images of the sample's microstructure were obtained with the help of a JSM-IT300 scanning electron microscope from JEOL (Tokyo, Japan). Equipping the microscope with a X-MAX80 attachment for X-ray microanalysis from Oxford Instruments (Abingdon, UK) allowed us to identify the contents of elements in individual phase formations, which can be seen on the microstructure images. The sample's microstructure was studied at an accelerating voltage of 10–25 kV and an electron probe current of 52–96 μA. The operating distance to the sample surface was 10.5–11.7 mm.

## 3. Results

As one can see in Figure 1, reaction 25, providing the reduction of molybdenum oxide $MoO_3$ to $MoO_2$ when reacting with carbon, had the highest probability in the Ni-Cr-Mo-W-O-C system within the studied temperature range of 300–2000 K. Curve 25 of this reaction is located in the negative part of the diagram.

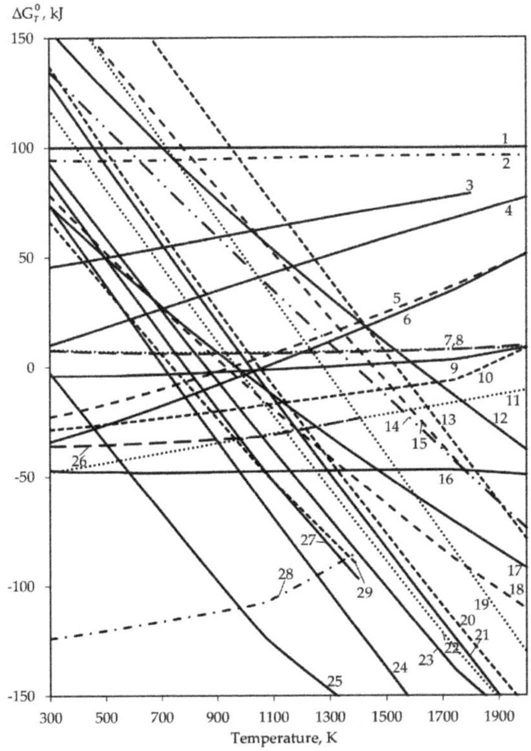


**Figure 1.** Dependence of the Gibbs free energy difference on the carbon reduction reaction temperature in the Ni-Cr-Mo-W-O-C system: 1—$7/33Cr_2O_3 + CO = 2/33Cr_7C_3 + 27/33CO_2$; 2—$1/3Cr_2O_3 + CO = 2/3Cr + CO_2$; 3—$23/93Cr_2O_3 + CO = 2/93Cr_{23}C_6 + 81/93CO_2$; 4—$3/17Cr_2O_3 + CO = 2/17Cr_3C_2 + 13/17CO_2$; 5—$1/3MoO_2 + CO = 1/6Mo_2C + 5/6CO_2$; 6—$1/5WO_3 + CO = 1/5WC + 4/5CO_2$; 7—$1/2MoO_2 + CO = 1/2Mo + CO_2$; 8—$1/2WO_2 + CO = 1/2W + CO_2$; 9—$1/3WO_3 + CO = 1/3W + CO_2$; 10—$WO_3 + CO = WO_2 + CO_2$; 11—$1/4MoO_3 + CO = 1/8Mo_2C + 7/8CO_2$; 12—$7/27Cr_2O_3 + C = 2/27Cr_7C_3 + 21/27CO$; 13—$1/3Cr_2O_3 + C = 2/3Cr + CO$; 14—$3/13Cr_2O_3 + C = 2/13Cr_3C_2 + 9/13CO$; 15—$23/81Cr_2O_3 + C = 2/81Cr_{23}C_6 + 69/81CO$; 16—$NiO + CO = Ni + CO_2$; 17—$1/3WO_2 + C = 1/3WC + 2/3CO$; 18—$1/4WO_3 + C = 1/4WC + 3/4CO$; 19—$2/5MoO_2 + C = 1/5Mo_2C + 4/5CO$; 20—$1/2MoO_2 + C = 1/2Mo + CO$; 21—$1/2WO_2 + C = 1/2W + CO$; 22—$1/3WO_3 + C = 1/3W + CO$; 23—$WO_3 + C = WO_2 + CO$; 24—$NiO + C = Ni + CO$; 25—$MoO_3 + C = MoO_2 + CO$; 26—$1/3MoO_3 + CO = 1/3Mo + CO_2$; 27—$1/3MoO_3 + C = 1/3Mo + CO$; 28—$MoO_3 + CO = MoO_2 + CO_2$; 29—$2/7MoO_3 + C = 1/7Mo_2C + 6/7CO$.

When the temperature rises above 700 K, the reduction reaction of nickel oxide NiO to Ni (curve 24) takes negative values of the Gibbs free energy difference. At a temperature above 950 K, the negative values of the Gibbs free energy difference are taken by the reaction of the molybdenum trioxide reduction to $Mo_2C$ carbide, and above 800 K to Mo. Reaction 23, which is typical for the reduction of $WO_3$ to $WO_2$, which takes negative values of the Gibbs free energy difference at a temperature above 850 K. In the temperature range of 920–1050 K, the negative values of the Gibbs free energy difference are taken by the reactions of tungsten oxides $WO_3$ (reactions 18, 22) and $WO_2$ (reactions 17, 21) reductions to WC and W, respectively (Figure 1). In the case of $MoO_2$, one can observe reactions 20 and 19 with the formation of $Mo_2C$ and Mo, respectively.

In the temperature range of 1400–1600 K, the reactions involving the carbonothermal reduction of chromium oxide $Cr_2O_3$ (reactions 12–15) take negative values of the Gibbs free energy difference. Reactions with the formation of $Cr_3C_2$ and $Cr_{23}C_6$ carbides can most likely be seen from the thermodynamic point of view (reactions 14 and 15, respectively). The reduction reaction of chromium oxide $Cr_2O_3$ to carbide $Cr_7C_3$ had the lowest thermodynamic probability; thus, the relevant curve is seen in the negative part of the diagram at a temperature of 1600 K.

The oxide reduction reaction curves of the discussed system with carbon monoxide show the lowest values of the Gibbs free energy difference in areas of relatively low temperatures. However, regarding the thermodynamic probability of an oxide reduction reaction, a clear analogy with the interaction with carbon can be seen. This means that the probability of the occurrence of a reduction reaction for the relevant oxides decreases in the following sequence: NiO, $MoO_3$, $WO_3$, $MoO_2$, $WO_2$, $Cr_2O_3$. It should also be noted that reactions 11, 16, 26, and 28 can be found in the negative part of the diagram, and reactions 9 and 10 fall within the positive part of the diagram only at temperatures of 1400 K and 1800 K, respectively. The curves of reactions 1–4, 7, and 8 occur in the positive part of the diagram. Reactions 1 and 2 of $Cr_2O_3$ reduction to $Cr_7C_3$ and Cr, respectively, have the highest value of Gibbs free energy difference.

Studies of the phase composition of the obtained alloying compound showed relatively active development of the FeNi phase, as well as FeCr (Figure 2). The carbide component was expressed as the development of $Fe_3C$, tungsten carbides WC and $W_2C$, and complex carbide $W_2C{\cdot}Mo_2C$.

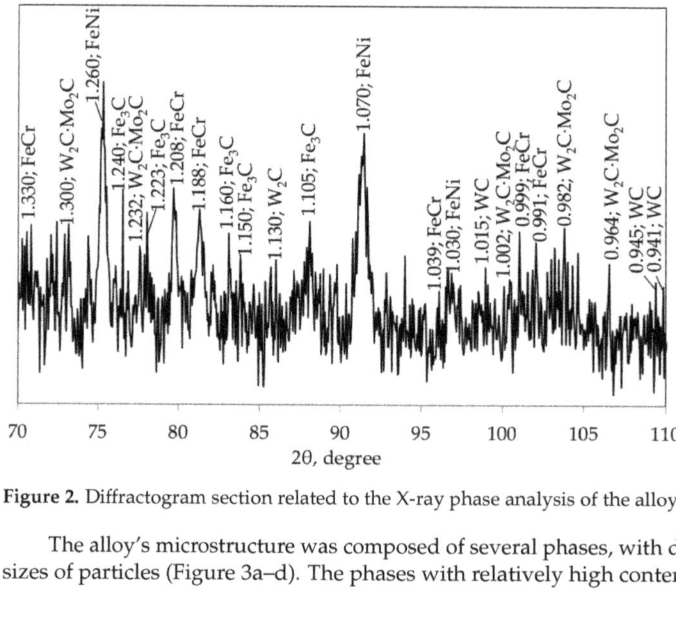


**Figure 2.** Diffractogram section related to the X-ray phase analysis of the alloying compound.

The alloy's microstructure was composed of several phases, with different shapes and sizes of particles (Figure 3a–d). The phases with relatively high contents of Ni (areas 3, 7,

9, and 15) and Cr (areas 5, 8, 11, 12, and 16), up to 44.33 wt% and 46.33 wt%, respectively, were identified in the studied areas (Figures 3 and 4, Table 1). Some particles had high contents of Mo and W (areas 2, 4, 10, 13), up to 14.17 wt% and 26.52 wt%, respectively. These particles also contained Nb up to 1.59 wt%. The carbon content in the above areas (Figures 3 and 4, Table 1) varied within a range of 0.45–2.26 wt%. Moreover, some phase formations with carbon contents up to 89.54 wt% were found, and they seemed to be the residual unreacted carbon-reducing agents.

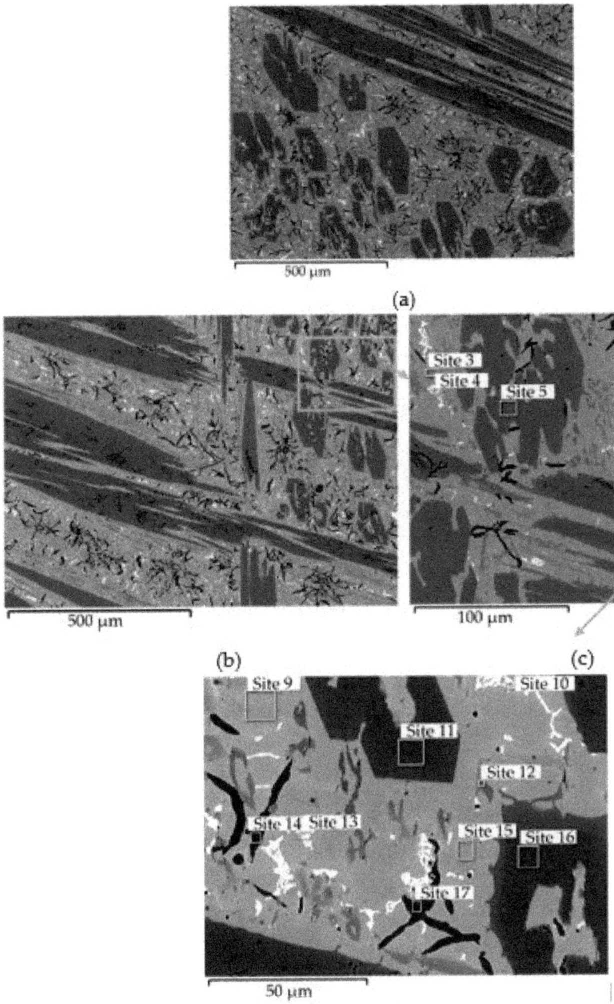


**Figure 3.** Images of the alloying compound's microstructure, with magnifications of: (**a**,**b**)—×100, (**c**)—×500, (**d**)—×1000, 1–17—X-ray microanalysis areas. Arrows indicate scaled points of microstructure.

The determination of element concentrations in the studied area of the sample surface showed a clear distribution between the phases with different contents of Cr and Ni (Figure 5c,d). The Fe content was relatively uniform throughout the studied area (Figure 5b), except for carbon inclusions and phases with high Mo and W contents. However, the highest concentration of iron was observed in the phase with a high nickel content, which is evidenced in the figure by the corresponding more intensively colored areas. W and Mo, as well as Nb, were more concentrated in separate inclusions (Figure 5e–g).

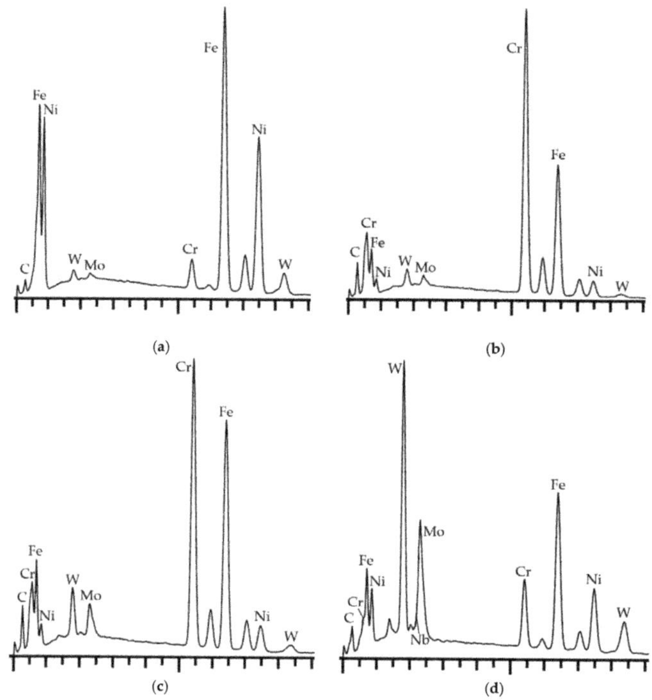


**Figure 4.** Spectrographs of the X-ray microanalysis of some studied areas of the alloy obtained in Figure 3, respectively: (**a**)—9; (**b**)—11; (**c**)—12; (**d**)—13.

**Table 1.** Results of the X-ray microanalysis of the alloying compound, based on Figure 3.

| Areas Studied | Element Content, wt% | | | | | | | | | |
|---|---|---|---|---|---|---|---|---|---|---|
| | C | Al | Cr | Mn | Fe | Ni | Nb | Mo | W | Total |
| 1 | 81.37 | 0.00 | 0.87 | 0.00 | 9.92 | 7.84 | 0.00 | 0.00 | 0.00 | 100 |
| 2 | 2.26 | 0.15 | 5.46 | 0.00 | 41.50 | 22.13 | 0.85 | 9.79 | 17.86 | 100 |
| 3 | 0.47 | 0.07 | 3.77 | 0.11 | 48.97 | 44.33 | 0.00 | 0.41 | 1.87 | 100 |
| 4 | 2.21 | 0.07 | 5.91 | 0.00 | 40.28 | 23.03 | 1.02 | 9.18 | 18.30 | 100 |
| 5 | 1.81 | 0.08 | 46.33 | 0.00 | 42.90 | 5.42 | 0.00 | 1.09 | 2.37 | 100 |
| 6 | 85.53 | 0.00 | 0.98 | 0.00 | 7.67 | 5.82 | 0.00 | 0.00 | 0.00 | 100 |
| 7 | 0.45 | 0.10 | 3.62 | 0.09 | 49.52 | 44.09 | 0.00 | 0.39 | 1.74 | 100 |
| 8 | 1.85 | 0.06 | 46.17 | 0.00 | 42.67 | 5.65 | 0.00 | 1.16 | 2.44 | 100 |
| 9 | 0.48 | 0.00 | 3.16 | 0.10 | 51.61 | 41.53 | 0.00 | 0.67 | 2.45 | 100 |
| 10 | 2.25 | 0.12 | 7.26 | 0.00 | 36.27 | 15.51 | 1.44 | 13.3 | 23.85 | 100 |
| 11 | 1.82 | 0.03 | 45.10 | 0.00 | 43.39 | 5.60 | 0.00 | 1.36 | 2.70 | 100 |
| 12 | 1.89 | 0.06 | 31.97 | 0.00 | 50.52 | 6.56 | 0.00 | 3.52 | 5.48 | 100 |
| 13 | 2.23 | 0.17 | 7.78 | 0.00 | 33.07 | 14.47 | 1.59 | 14.17 | 26.52 | 100 |
| 14 | 89.54 | 0.00 | 1.09 | 0.00 | 6.02 | 3.35 | 0.00 | 0.00 | 0.00 | 100 |
| 15 | 0.46 | 0.11 | 3.70 | 0.08 | 50.02 | 43.75 | 0.00 | 0.31 | 1.57 | 100 |
| 16 | 1.84 | 0.05 | 45.46 | 0.00 | 43.17 | 5.54 | 0.00 | 1.23 | 2.71 | 100 |
| 17 | 84.02 | 0.00 | 0.71 | 0.00 | 8.94 | 6.33 | 0.00 | 0.00 | 0.00 | 100 |

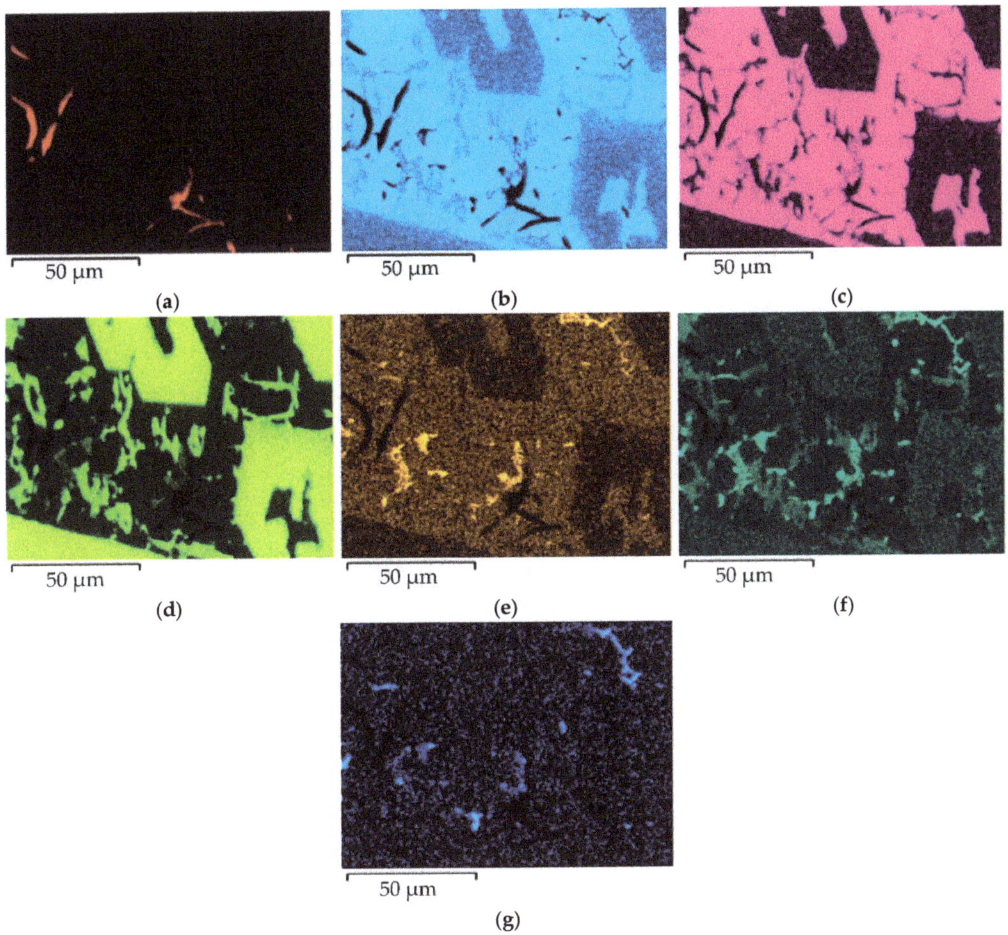


**Figure 5.** The results of X-ray microanalysis in the major elements distribution mode in the characteristic X-rays, according to Figure 3d (magnification ×1000). The higher element contents correspond to more intense colors: **(a)**—C, **(b)**—Fe, **(c)**—Ni, **(d)**—Cr, **(e)**—Mo, **(f)**—W, **(g)**—Nb.

## 4. Discussion

It follows from the results obtained from thermodynamic analysis in the Ni-Cr-Mo-W-O-C system regarding the reduction of metal oxide technogenic raw materials that the stability of oxide compounds of refractory elements decreases as the temperature rises. From a thermodynamic point of view, higher oxide compounds have less stability than lower oxide ones. In the system which we studied, the formation of metal oxides and carbides, free metals, and CO and $CO_2$ are possible. The refractory metal oxides are reduced by carbon and carbon monoxide. A high probability of parallel carbide formation reactions, along with the reduction reactions at relatively low temperatures, has been discovered. As the temperature rises, the probability of metal reduction in a free state increases. This additionally dictates a negligible probability of obtaining a carbon-free product. The consideration of phase transitions and the change in the heat capacity of the reaction components along with the temperature increase allowed us to obtain more accurate results. The significant number of concerned reactions in a wide temperature range demonstrates a broader picture of possible transformations in the process of alloyed technogenic waste reduction. Difficulties in the development of this area lie in the lack of a sufficiently large reference base to be used for calculations. There are possible prospects of studies in this direction for calculating, plotting, and updating the equilibrium reaction diagrams in the Ni-Cr-Mo-W-O-C system with the identification of the formed areas. In addition, it will be possible to establish the most acceptable conditions for the existence of phases depending on changes in the temperature and partial pressure of CO. The plotted diagram can be used in the technological field for the purpose of obtaining metallized materials from alloyed steel waste.

The X-ray phase analysis of the studied compound (Figure 2) shows that the alloying elements Cr and Ni mostly occurred in the form of intermetallic phases FeNi and FeCr, respectively. Some alloying elements could be found in the form of substitutional atoms in $Fe_3C$. This agrees with the results of the study cited in [9], where the diffractogram of the reduced alloyed product clearly showed only iron-containing compounds, which is also typical for unalloyed raw materials [8]. However, in contrast to the above study, the completed studies also showed some carbide compounds of the alloying elements, such as WC, $W_2C$, and $W_2C{\cdot}Mo_2C$. This agrees with the results of [12] regarding the parallel reduction and carbide formation and the practical impossibility of obtaining a completely carbon-free product, as well as with the results of [7]. In this case, the residual carbon in the form of carbides will further ensure a relatively high reducibility when the alloying element is used.

The studies of the microstructure, together with the X-ray microanalysis of the obtained alloy, further demonstrate the nature of the elements' presence within it. The inclusions with higher contents of nickel with iron (Figure 3c,d, areas 3, 7, 9, 15) appear to constitute the intermetallic phase of FeNi, which is evidenced by the completed X-ray phase analyses (Figure 2). This agrees well with the results of [15], where the relatively high efficiency of the carbon reduction of the iron–nickel oxide component was confirmed. At the same time, [15] showed that the iron-containing carbide component Fe ($Fe_nC_m$) was present together with $Fe_xNi_y$ in the course of the reduction process. Some phases were characterized by the high content of Cr and Fe (Figure 3c,d, areas 5, 8, 11, 12, 16). These appear to be the intermetallic FeCr phases which were seen on the obtained diffractogram (Figure 2). Taking into account the results of the X-ray phase analyses, the above areas can contain $Fe_3C$ carbide, where Fe atoms are partially substituted by Cr atoms. This agrees with the completed thermodynamic studies and the results of [11,13,14], which demonstrated the formation of chromium-containing carbides in parallel with the reduction reactions. However, in [13,14], the reduction was accompanied by the formation of separate chromium carbides of the $Cr_nC_m$ type. The absence of chromium carbides in the diffractogram in these studies can be explained by the fact that some carbides can be dissolved in the intermetallic component FeCr when two phases come into contact. Similar processes are described in [11], where the formed chromium carbides were dissolved in $\gamma$-Fe. In turn, the thermodynamic studies (Figure 1) showed a lack of chromium carbides and metallic chromium, along with a relatively low probability of the corresponding reactions.

Some particles with higher contents of Mo, W, Nb, and C (Figure 3, areas 2, 4, 10, 13) can constitute the carbide phases. This corresponds to the results of the thermodynamic (Figure 1) and phase (Figure 2) studies, based on which we determined that the phases in the specified surface areas may contain carbides such as WC, $W_2C$, and $W_2C{\cdot}Mo_2C$. This also agrees with the results of studies regarding the reduction mechanisms of W [17–19] and Mo [16] oxides. It was also found that the metallic and carbide components were formed in parallel at the final stage of the reduction. Here, as a contrast, the assumed particles of the carbide phase in these studies were characterized by high contents of the alloying element complexes. This indicates a more complex nature of the refractory element compounds. Analysis of the study results indicates the predominance of intermetallic phase formations of iron with nickel and chromium in the phase composition. The fraction of residual carbon in the form of a carbide component had a relatively low presence, thus ensuring that the reducing ability is required when the alloy is used.

The limitations of the study's results encourage the use of the obtained alloying composite in relation to certain steel grades, based on the set of present elements. For example, the complex of elements Ni, Cr, Mo, and W in some heat-resistant steels identifies the required qualitative properties of the product. In contrast to Cr, Mo, and W, the Ni content in high-speed tool steels is highly limited to tenths of a percent. Therefore, the allowable limits of the element content can be exceeded in the target product. Similar problems cannot be excluded for other steel grades where there are severe limits on the content of one or more elements in the alloying compound. Thus, to avoid problems of this nature and to increase the use rates of raw materials, close ratios of the element contents in the alloying composite and in the target product should be followed.

The lack of results of X-ray analyses for several phase formations at once, which would have allowed us to obtain the averaged contents of the elements, should be noted as a disadvantage. This would make it possible to monitor the distribution of the major elements throughout the studied surface of the samples, which would characterize, to a greater degree, the nature of the phases and inclusions.

This study can be continued in order to expand the range of the steel grades, the oxide wastes of which can be involved in processing by reduction smelting. The difficulties in continuing this study are due to the lack of a sufficient experimental database. The most promising for prospective studies are wastes with high levels of alloying elements.

The completed studies determined new technological aspects involved in the processing of high-alloy technogenic waste to obtain compounds with relatively low residual carbon content. The resulting parameters of the resource-saving alloying compound can make it possible to replace some standard ferroalloys of those steel grades in the manufacturing process that place limitations on the carbon content. Heat-resistant austenitic steels that are melted in electric arc furnaces are promising in these terms. The obtained alloy did not contain compounds and phases with relatively high tendencies for sublimation. That is, there is no need to create special conditions to prevent evaporation and losses of alloying elements with the gas phase. This also leads to wider use of the alloying elements.

## 5. Conclusions

1. The features of thermodynamic equilibrium in the Ni-Cr-Mo-W-O-C system in relation to the reduction of oxide technogenic raw materials were determined. This provided a satisfactory notion of the possibility of the reactions, depending on the temperature. The thermal conditions that can ensure the high probability of reactions forming carbides and free elements during carbon and carbon monoxide reduction were clarified. The probability of reduction reactions of the corresponding oxides decreases in the following order: nickel oxide, trioxides, molybdenum and tungsten dioxides, and chromium oxide. The reduction reactions of higher oxides are more probable than those of lower oxides. Moreover, obtaining a reduced product without carbides is unlikely.

2. The features of the phase composition of the obtained alloying compound were determined. The intermetallic phases FeNi and FeCr were developed quite intensively. The phase of $Fe_3C$ carbide clearly occurred. The fragmentary development of the carbides of the alloying elements WC, WC, $W_2C$, and $W_2C{\cdot}Mo_2C$ was found. The absence of compounds and phases with relatively high tendencies for sublimation was determined. That is, there is no need to create special conditions in order to prevent the evaporation and losses of alloying elements with the gas phase, which leads to the wider use of the alloying elements.

3. It was found that the alloy microstructure was composed of several phases with particles of different shapes and sizes. Inclusions with relatively high contents of iron and nickel (up to 44.33 wt%), as well as chromium (46.33 wt%), which were found in the studied areas appeared to be of intermetallic nature. Moreover, the phase formations with carbon contents up to 89.54 wt% were found, and they appeared to be the residual unreacted carbon-reducing agents. Some particles had higher contents of Mo and W, up to 14.17 wt% and 26.52 wt%, respectively, and of Nb, up to 1.59 wt%. Such inclusions were composed mainly of iron carbides and alloying elements.

**Author Contributions:** Conceptualization, A.P. and T.B.; methodology, A.A. and T.B.; software, A.P. and A.A.; validation, A.S. (Aleksey Smirnov) and A.A.; formal analysis, A.S. (Anastasiia Semenko) and Y.S.; investigation, A.P.; resources, A.S. (Anastasiia Semenko) and Y.S; data curation, A.A. and T.B.; writing—original draft preparation, A.P.; writing—review and editing, A.S. (Aleksey Smirnov) and T.B.; visualization, A.A. and A.P.; supervision, A.A.; project administration, A.P. All authors have read and agreed to the published version of the manuscript.

**Funding:** This research received no external funding.

**Data Availability Statement:** Not applicable.

**Conflicts of Interest:** The authors declare no conflict of interest.

## References

1. Henckens, M.L.C.M.; Van Ierland, E.C.; Driessen, P.P.J.; Worrell, E. Mineral resources: Geological scarcity, market price trends, and future generations. *Resour. Policy* **2016**, *49*, 102–111. [CrossRef]
2. Puchol, R.Q.; Sosa, E.R.; González, L.O.; Castañeda, Y.P.; Sierra, L.Y. New conception of the reutilization of solid waste from Cuban nickeliferous hydrometallurgical industry. *Centro Azucar* **2016**, *43*, 1–15.
3. Grigoriev, S. Some tactical activities increase efficiency in resource metallurgy of special steels. *State Reg. Ser. Econ. Entrep.* **2014**, *1*, 68–72.
4. Sekiguchi, N. Trade specialisation patterns in major steelmaking economies: The role of advanced economies and the implications for rapid growth in emerging market and developing economies in the global steel market. *Miner. Econ.* **2017**, *30*, 207–227. [CrossRef]
5. Maksimov, E.A.; Vasil'ev, V.I. The utilization of the wastes of the rolling and pipe-rolling shops of the metallurgical plants and their processing. ferrous metallurgy. *Bull. Sci. Tech. Econ. Inf.* **2016**, *3*, 99–107.
6. Grigor'ev, S.M.; Petrishchev, A.S. Assessing the phase and structural features of the scale on P6M5ф3 and P12M3K5ф2 steel. *Steel Transl.* **2012**, *42*, 272–275. [CrossRef]
7. Grigoriev, D.S.; Skachkov, V.O.; Karpenko, G.V. Methodological principles for evaluating the economic efficiency of the utilization of rare metals from technogenic waste of special steels. *Metallurgy* **2022**, *1*, 87–92. [CrossRef]
8. Mechachti, S.; Benchiheub, O.; Serrai, S.; Shalabi, M. Preparation of iron powders by reduction of rolling mill scale. *Int. J. Sci. Eng. Res.* **2013**, *4*, 1467–1472.
9. Petryshchev, A.; Milko, D.; Borysov, V.; Tsymbal, B.; Hevko, I.; Borysova, S.; Semenchuk, A. Studying the physical-chemical transformations at resource-saving reduction melting of chrome-nickel-containing metallurgical waste. *East.-Eur. J. Enterp. Technol.* **2019**, *2*, 59–64. [CrossRef]
10. Hryhoriev, S.; Bikulov, D.; Skachkov, V.; Berezhnaya, O.; Oliynyk, O. Intensification of the process of metallizing tungsten containing ore raw materials by the powder metallurgy method. *East.-Eur. J. Enterp. Technol.* **2019**, *4*, 31–36. [CrossRef]
11. Zhao, L.; Wang, L.; Chen, D.; Zhao, H.; Liu, Y.; Qi, T. Behaviors of vanadium and chromium in coal-based direct reduction of high-chromium vanadium-bearing titanomagnetite concentrates followed by magnetic separation. *Trans. Nonferrous Met. Soc. China* **2015**, *25*, 1325–1333. [CrossRef]
12. Ryabchikov, I.V.; Belov, B.F.; Mizin, V.G. Reactions of metal oxides with carbon. *Steel Transl.* **2014**, *44*, 368–373. [CrossRef]
13. Hryhoriev, S.; Petryshchev, A.; Shyshkanova, G.; Yakimtsov, Y.; Zhuravel, S.; Yamshinskij, M.; Fedorov, G.; Kruglyak, D.; Shevchenko, O.; Goliev, Y. Study into properties of the resource saving chromium containing briquetted alloying additive from ore raw materials. *East.-Eur. J. Enterp. Technol.* **2017**, *4*, 38–43. [CrossRef]
14. Simonov, V.K.; Grishin, A.M. Thermodynamic analysis and the mechanism of the solid-phase reduction of $Cr_2O_3$ with carbon: Part 1. *Russ. Metall. (Metally)* **2013**, *6*, 425–429. [CrossRef]
15. Zhang, Y.; Wei, W.; Yang, X.; Wei, F. Reduction of Fe and Ni in Fe–Ni–O systems. *J. Min. Metall. Sect. B Metall.* **2013**, *49*, 13–20. [CrossRef]
16. Zhu, H.Y.; Li, Z.B.; Yang, H.S.; Luo, L.G. Carbothermic reduction of $MoO_3$ for direct alloying process. *J. Iron Steel Res. Int.* **2013**, *20*, 51–56. [CrossRef]
17. Kozyrev, N.A.; Galevsky, G.V.; Valuev, D.V.; Shurupov, V.M.; Kozyreva, O.E. Surfacing with tungsten-containing ores. *IOP Conf. Ser. Mater. Sci. Eng.* **2015**, *91*, 012009. [CrossRef]
18. Shveikin, G.P.; Kedin, N.A. Products of carbothermal reduction of tungsten oxides in argon flow. *Russ. J. Inorg. Chem.* **2014**, *59*, 153–158. [CrossRef]
19. Islam, M.; Martinez-Duarte, R. A sustainable approach for tungsten carbide synthesis using renewable biopolymers. *Ceramics International* **2017**, *43*, 10546–10553. [CrossRef]
20. Alekseev, A.; Bovkun, G.; Bolgar, A. *Properties, Preparation and Use of Refractory Compounds*, 1st ed.; Metallurgy: Moscow, Russia, 1986; 928p.
21. Kireev, V. *Methods of Practical Calculations in the Thermodynamics of Chemical Reactions*; Chemistry: Moscow, Russia, 1970; 328p.
22. Kazachkov, E. *Calculations on the Theory of Metallurgical Processes*; Metallurgy: Moscow, Russia, 1988; 288p.
23. Samsonov, G. *Refractory Compounds*; Metallurgy: Moscow, Russia, 1976; 560p.

24. Elyutin, V. *High-Temperature Materials, Part 1*; Metallurgy: Moscow, Russia, 1972; 264p.
25. Gurvich, L.; Veits, I.; Medvedev, V. *Thermodynamic Properties of Individual Substances. Reference Edition: In 4 Volumes vol. T. IV. Book 1*; Nauka: Moscow, Russia, 1982; 623p.

minerals

MDPI

*Article*

# Prevention of Silica Gel Formation for Eudialyte Study Using New Digestion Reactor

**Ivan Silin [1], Christian Dertmann [2], Vesna S. Cvetković [3], Srecko Stopic [2,*] and Bernd Friedrich [2]**

1 Unit of Mineral Processing, RWTH Aachen University, 52064 Aachen, Germany; silin@amr.rwth-aachen.de
2 IME Process Metallurgy and Metal Recycling, RWTH Aachen University, 52056 Aachen, Germany; cdertmann@ime-aachen.de (C.D.); bfriedrich@ime-aachen.de (B.F.)
3 Department of Electrochemistry, Institute of Chemistry, Technology and Metallurgy, National Institute, University of Belgrade, Njegoševa 12, 11000 Belgrade, Serbia; v.cvetkovic@ihtm.bg.ac.rs
* Correspondence: sstopic@ime-aachen.de; Tel.: +49-241-80-95860

**Abstract:** This work includes a combined hydrometallurgical treatment of a eudialyte ore sample with a subsequent preparation of leaching residue using mechanical separation methods. Hydrometallurgical treatment contains dry digestion with sulphuric acid at room temperature and filtration of the obtained product. The objectives of adopting these procedures are to test a new digestion reactor in order to prevent silica gel formation from the eudialyte ore. The obtained results revealed that silica gel formation is prevented during dissolution with sulphuric acid. A high leaching efficiency of light rare earth elements (La, Ce, Nd, Y) was reached using the dry digestion process with sulphuric acid, where the starting molarity was 12 mol/L. After the filtration process, magnetic separation is studied as the main method to recover weakly magnetic minerals like amphiboles and pyroxenes from the leaching residue in the magnetic fraction and feldspars in the nonmagnetic fraction. A new combined research strategy was developed for the production of different concentrates such as the one bearing Zr, Hf, and Nb.

**Keywords:** hydrometallurgy; dry digestion; silica gel prevention; magnetic separation; eudialyte; aegirine; feldspar

## 1. Introduction

Eudialyte is a unique mineral and nowadays is known as a potential source of rare earth and high-field-strength elements in the absence of radioactive elements such as uranium and thorium, which is the case for bastnasite, xenotime, monazite and steenstrupine. Significant resources of eudialyte and accompanied zirconosilicates are represented by Ilimaussaq and Tanbreez (Greenland), Norra Kärr (Sweden), Khibina and Lovozero (Russia), Mont Saint-Hilaire and Kipawa (Canada), Pilanesberg (South Africa), Saima (China) and Poços de Caldas (Brazil) [1,2]. Besides eudialyte and accompanied complex Zr and Ti silicates, the eudialyte ore consists of mafic minerals like amphiboles and pyroxenes and of alkali feldspars and feldspathoids like nepheline [3]. According to the current state of the art, dry and wet high-intensity magnetic separation (WHIMS) is widely proposed as the main method to recover weakly magnetic minerals with amphiboles, pyroxenes and eudialyte for the magnetic fraction, and feldspars and nepheline for the non-magnetic fraction [4]. Moreover, various marketable products could be additionally obtained from eudialyte ores as by-products like nepheline-syenite and feldspars.

A study on the extraction of REE as critical metals from eudialyte has been mainly confined to laboratory studies and combined hydrometallurgical treatments, producing different precipitation products such as rare earth carbonate and rare earth oxalate. Despite being easily dissolved by acids such as hydrochloric acid and sulphuric acid, the main challenge with this processing has been how to avoid silica gel formation, which is a gelatinous and non-filterable phase [5]. Lebedev et al. [6,7] have studied a two-stage decomposition process for REE recovery. High-temperature leaching with concentrated sulphuric acid ($H_2SO_4$) was used to decompose the eudialyte. Then, following washing of the slurry with sodium sulphate solution, the REE remained in the insoluble residue as double sulphate salts. The residue was then washed again and REE recovered after converting the sulphates into nitrates or chlorides. Ma et al. [8] have studied the effect of digestion time on REE extraction using an acid/concentrate ratio of 1.25:1. The treated concentrate was subsequently leached at room temperature and a water/concentrate ratio of 2:1 for 1 h. The digestion process was performed without external heating, but the reaction temperature reached 70–80 °C because of exothermic effects. Another idea for avoiding gel formation is the introduction of sodium fluoride, because fluoride ions ($F^-$) can promote the coagulation of silica acid in solution such that silica content in the pregnant would be lower and the slurry thus filterable [9]. $F^-$ ions can promote eudialyte decomposition, but this is also toxic, and it can destroy glass equipment.

The valorisation of various waste is very important for the sustainable development of a green and circular economy [10–14]. Therefore, this study proposes a mechanical separation of valuable mineral products from leaching residues with the goal being to minimise the volume of the tailings storage. The most valuable minerals of the leaching residues are feldspars and mafic minerals, represented by pyroxenes like aegirine-augite groups. Due to the differences in the specific magnetic susceptibility of aegirine-augite groups and feldspars, the magnetic separation could also be considered as the main method for the treatment of solid leaching residues [15,16]. Feldspars are used in the glass and ceramic industries as filling materials [17,18]. Aegirine could be used as pigments, mineral wools, refractory, ceramic and glass industries (glazes, black glass, decorative glassware), fillers, coatings, lubricants for hot metal forming, weighting agents or as low-melting flux for pyrometallurgical refinement of copper from impurities [19–21].

The first aim of this study is the extraction of rare earth elements using dry digestion process in special constructed reactor preventing the formation silica gel formation from a eudialyte ore sample from Norra Kärr deposit. This digestion reactor was developed from a 1 L laboratory reactor to an increased 100 L one at the Institute for Process metallurgy and Metal recycling of the RWTH Aachen University in Germany [22,23] in contrast to some bioleaching processes [24], where the extraction efficiency for the single REEs was a maximum of 35% for zircon concentrates but up to about 70% for the eudialyte concentrate. A large number of original research articles covering techniques including solvent extraction, ion exchange, precipitation and other emerging technologies have been reviewed in detail and critically assessed to understand the role and behaviour of specific impurities in REE processing (including Al, Ca, Mg, Fe, Si, Th, U, Ti, Zr, Hf, Cr, Mo, Mn, Co, Ni, Cu, Zn, Sn, Pb and Bi) [25]. The second aim of this work is to is to investigate the possibility of producing valuable mineral products like aegirine and feldspar products from solid leaching residue using mechanical separation methods in the final step.

## 2. Materials and Methods

### 2.1. Materials

The original sample of low-grade eudialyte ore from the Norra Kärr deposit in Sweden was used for various geometallurgical investigations during the multidisciplinary Siemens research section (S-FB): Rare Earth—Green Mining and Separation. The remaining processing products from geometallurgical investigations (sensor-based sorting, magnetic and electrical separation, flotation) were mixed, homogenised and split into subsamples for hydrometallurgical investigations in this work. Initially, the eudialyte ore sample was crushed by a jaw crusher (Type MN 931/10. KHD Humboldt Wedag AG, Cologne, Germany) and then ground by a roll crusher (Type LWBP 2/2, Karl Merz Maschinenfabrik GmbH, Heschingen, Germany) in a circuit with a 180 µm sieve. After hydrometallurgical treatment in new digestion reactor, the pulp was filtered using a filter press to obtain the filter cake (leaching residue) and the filtrate (REE-bearing solution). The leaching residue was then used for magnetic separation tests. The experimental process flowsheet is given in Figure 1.

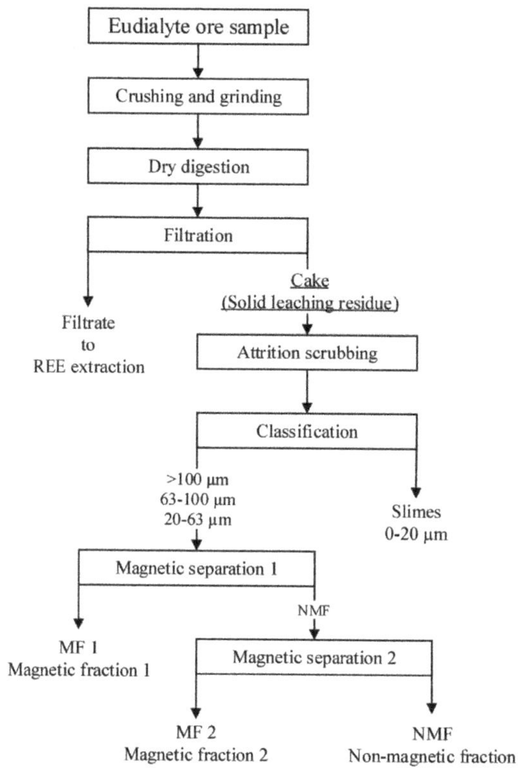


**Figure 1.** Experimental process flowsheet.

The chemical composition of the low-grade eudialyte ore sample is given in Table 1.

**Table 1.** The chemical composition of eudialyte ore sample (wt.%).

| Si | Na | K | Al | Zr | Ti | Ca | Fe | Mn | P | LREE (La–Sm) | HREE (Y, Eu–Lu) |
|---|---|---|---|---|---|---|---|---|---|---|---|
| 25.15 | 6.85 | 3.16 | 8.65 | 0.79 | 0.12 | 1.31 | 3.27 | 0.14 | 0.045 | 0.2823 | 0.1514 |

The molar proportion of (Na + K)/Al ratio in the investigated sample is 1.15 (>1). The sample of investigated eudialyte ore belongs to agpaitic rocks. About 35% of REE belong to HREE. Zr and REE are mainly distributed in eudialyte. According to the mineralogical analysis, the eudialyte ore sample consists of eudialyte (9%), feldspars (50%), mafic minerals (16%) like aegirine-augite series and feldspathoids (23%) like nepheline and natrolite. The micas minerals (biotite and muscovite) are detected as trace phases with a content below 2%. The metal distribution of sieve fractions is shown in Figure 2.

The particle size distribution of the milled eudialyte ore sample shows that 80% of the particles are smaller than 120 µm. Nearly 30% and 15% of the particles are less than 63 and 20 µm, respectively. Over 70% of Zr and REE in the eudialyte ore sample is distributed in the fraction below 63 µm. Sulphuric acid, 95.0%–98.0%, 258205-2.5 L, Sigma-Aldrich Chemie GmbH, Schnelldorf, Germany, for dry digestion process. Tap water was used for leaching process after dry digestion process.

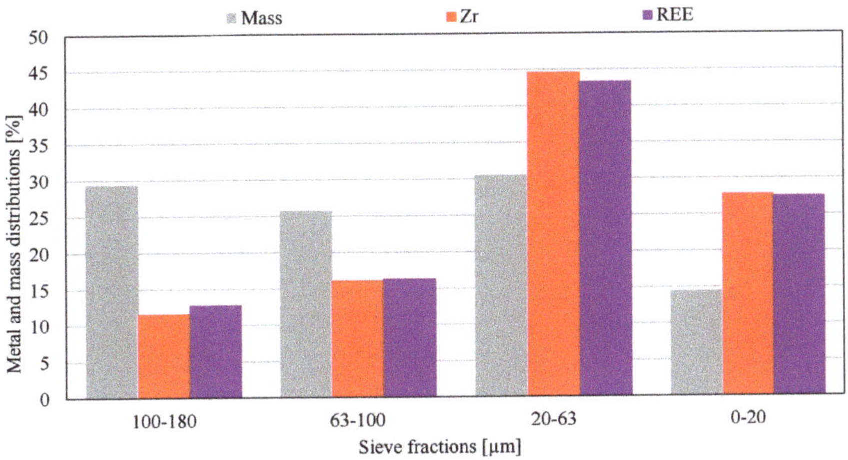


**Figure 2.** Particle size distribution and distribution of metals in sieve fractions of the eudialyte ore sample.

## 2.2. Methods

### 2.2.1. Sample Characterisation

The eudialyte ore and leaching residue were physically and chemically characterised. X-ray diffraction analyses of the eudialyte ore, solid leaching residue and mineral products were performed with an X-ray diffractometer (D2 Phaser, Bruker AXS, Karlsruhe, Germany), equipped with DIFFRAC.EVA Software, Version 5.0 and DIFFRAC.TOPAS Software, Version 5.0 for mineral identification and quantification, respectively. The solid samples were analysed via XRF (Panalytical Axios 2005, Malvern PANanalytical GmbH, Kassel, Germany), while the composition of liquid samples was determined via inductively coupled plasma optical emission spectroscopy (ICP-OES, Spectro Arcos 2013, SPECTRO Analytical Instruments GmbH, Kleve, Germany). The wet sieve analysis was provided using a sieve shaker (ASM 400 Siebtechnik GmbH, Mülheim a.d. Ruhr, Germany) to obtain the particle size distribution. An optical visualisation of mineral products was performed with a digital microscope (Keyence VHX-5000, Osaka, Japan). A pH meter (PCE-PHD1, PCE Deutschland GmbH, Meschede, Germany) was used to measure the pH value during attrition scrubbing.

### 2.2.2. Hydrometallurgical Treatment

The development of a modular reactor for the dry digestion of high-silicate ores and concentrates to avoid gel formation using eudialyte ores, as an example, is being carried out by the Institute for Process Metallurgy and Metal Recycling at RWTH Aachen University and the konzept GmbH, Engineering Services, Düren, realised together. In this study, a new reactor is developed for a process that has already been confirmed on a laboratory scale to avoid gelling (dry digestion) during leaching. Injection in the reactor is possible up to a scale of 75 kg (100 L), as shown in Figure 3a.

Special nozzle systems were used for acid injection aiming at leaching of metals and water injection for a dilution of suspension and additional leaching, as shown in Figure 3b. Furthermore, it should be proven that the modular system technology is robust in terms of chemical resistance and reproducibility of the results achieved. The innovation compared to process alternatives lies in the complete avoidance of external and cost-intensive heating energy as well as the previously impossible implementation of the dry digestion process in large solution volumes. Before gel formation, it is possible to convert the silicate components into crystals, which are filterable and stable. These results shall enable the cost-effective production of rare earth oxides.

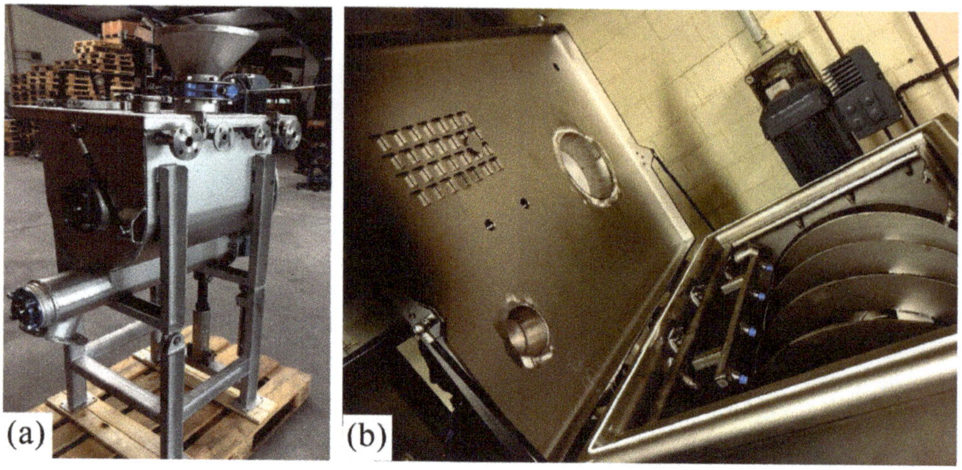

**Figure 3.** Dry digestion reactor (**a**); nozzle systems (blue) (**b**).

### 2.2.3. Magnetic Separation

Prior to the magnetic separation, the attrition scrubbing of leaching residue was provided in a laboratory-scale flotation machine (Type D-12, Denver Equipment Company, Littleton, CO, USA) in a 3 L beaker with 60% solids content. Impeller speed was set to 1600 rpm. Then, the treated leaching residue was sieved, dried and sampled for the magnetic separation tests, which is a widely used physical method to separate minerals with different magnetic properties. These tests were performed with a Frantz isodynamic magnetic separator (S.G. Frantz Company Inc., Tullytown, PA, USA), which is a device to accurately separate mineral grains according to their specific paramagnetic susceptibility. Three sieve fractions (>100, 63–100 and 20–63 μm) were used for the magnetic separation tests under various magnetic field strengths (0.5 and 1.6 T) at an orientation of magnetic system with a constant inclination angle (side slope of 15°, and forward tilt of 15°). As a result, two magnetic fractions (MF1 and MF2) and one non-magnetic fraction (NMF) were separated from each sieve fraction of 200 g.

## 3. Results and Discussion

### 3.1. Hydrometallurgical Treatment

The new dry digestor reactor, as shown in Figure 3, was used for hydrometallurgical treatment of eudialyte ore at room temperature using highly concentrated sulphuric acid. The value of 12 M is the starting molarity from the dry digestion stage. So, the 96% $H_2SO_4$ is diluted with water to achieve the desired 12 M. The tested experimental parameters for a treatment of eudialyte ore are shown in Table 2.

**Table 2.** Experimental parameters for hydrometallurgical treatment in dry digestor reactor.

| Parameter | Unit | Value |
|---|---|---|
| Initial Material | kg | 20 |
| Used acid ($H_2SO_4$) | % | 96 |
| Concentration | mol/L | 12 |
| Duration of dry digestion | min | 120 |
| Solid/liquid in dry digestion ($H_2SO_4$) | - | 1:2 |
| Solid/liquid in leaching ($H_2O$) | - | 1:5 |
| Duration of leaching | min | 60 |

The obtained results under these conditions are shown in Figure 4.

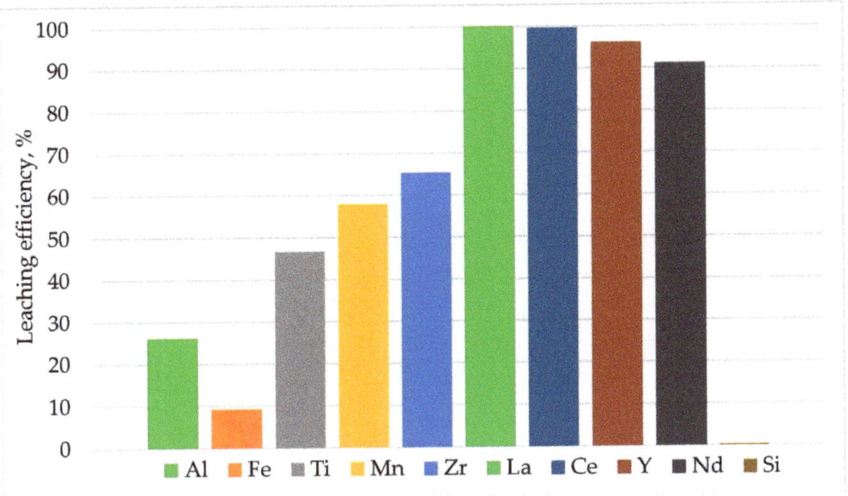


**Figure 4.** Leaching efficiency after hydrometallurgical treatment.

The obtained results have shown the high leaching efficiency (higher than 90%) at room temperature for rare earth elements (Nd, Y, Ce and La). At the same time, the leaching efficiency for aluminium and iron amounted to 10% and 25%, respectively. The minimal extraction achieved was for silicon (about 0.2%). This is the main reason that silica gel formation was prevented. The leaching efficiency for titanium, zirconium and magnesium amounted to between 45 and 65%., In principle, it is possible to receive higher Zr and, in the same turn, Hf yields. It is a question of the material composition and final pH in solution. As we lower the concentration of the sulphuric acid in the "washing" or "leaching" stage with a factor of 1:10, the final concentration is somewhat low compared to all the soluble elements in the input material. The final pH is about 2–3, and the Zr in the solution starts to precipitate. Therefore, if it is aimed to achieve higher Zr yields, it is possible to reach it starting with a higher acid molarity, change the s/l ratio or the dilution factor. But the main objective of this study is the extraction of the REE, and there is not high interest for this research, which is studied by Ma et al. [23]!

### *3.2. Magnetic Separation*

After filtration using filter press (ANDRITZ, Cologne, Germany) under working pressure approx. 5–6 bar below 50 °C, the filter cake (leaching residue) was sampled and wet sieved. Figure 5 shows the particle size distribution of the original filter cake.

According to the granulometric distribution in Figure 5, approximately 58% of the particles in the original filter cake are smaller than 100 µm. Preliminary visual examination of sieved fractions using the digital microscope shows that the mineral particles are cemented (Figure 6a) or coated (Figure 6b).

To liberate the aggregates of cemented mineral particles as well as to clean the mineral surfaces for further separation, the leaching residue was scrubbed in the flotation cell for 15 min and wet sieved. The pH value of pulp during the attrition scrubbing was measured at 3. After attrition scrubbing, the visual examination shows that the mineral particles were liberated and cleaned (Figure 6c). Moreover, after attrition scrubbing, the amount of slime particles smaller than 20 µm was raised from 14 up to 35%, while the amount of particles of over 100 µm was reduced from 42 to 15% (as shown in Figure 5). Due to the negative impact of slimes on physical separation processes, the fraction smaller than 20 µm was separated by wet sieving and was not further considered for the experiments in this work. The yield of the deslimed fraction is over 65%. The chemical composition of the leaching residue after desliming is listed in Table 3.

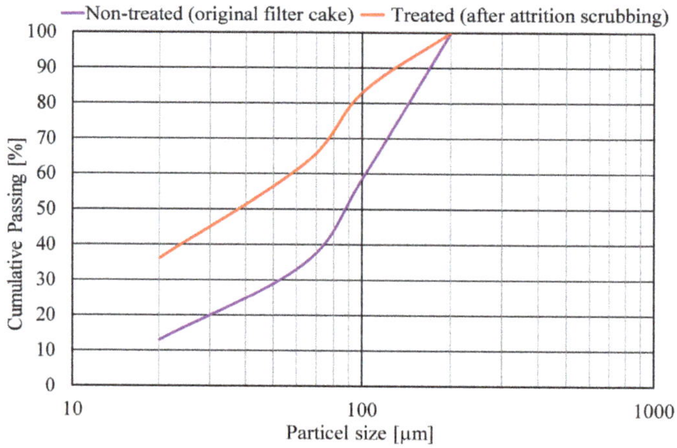


**Figure 5.** Particle size distribution of treated and non-treated leach residue.

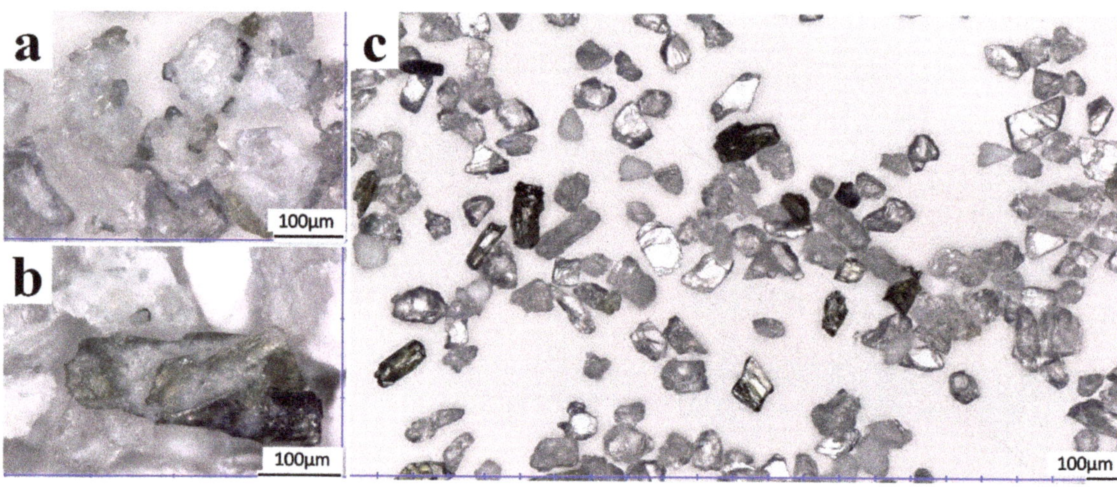


**Figure 6.** Optical microscope images of original filter cake (**a**,**b**) and treated leach residue (**c**).

**Table 3.** Chemical composition of leaching residue after desliming (%).

| $SiO_2$ | $Al_2O_3$ | $Na_2O$ | $K_2O$ | CaO | MgO | $Fe_2O_3$ | MnO | $ZrO_2$ | $TiO_2$ | $P_2O_5$ |
|---|---|---|---|---|---|---|---|---|---|---|
| 60.6 | 15.15 | 7.2 | 4.58 | 0.77 | 0.45 | 4.82 | 0.16 | 0.27 | 0.092 | 0.31 |

The main components in the leaching residue are $SiO_2$ and $Al_2O_3$. Fe, Ti and Mn mostly belong to iron-bearing silicates like pyroxenes and amphiboles and are partly distributed in the leached eudialyte phase. Zr also belongs to leached eudialyte. Feldspars are the major minerals remaining in the leach residue. The results of the magnetic separation are listed in Table 4.

**Table 4.** Results of magnetic separation.

| Products | Yield, % | $Fe_2O_3$ | | $ZrO_2$ | | $TiO_2$ | |
|---|---|---|---|---|---|---|---|
| | | Content, % | Recovery, % | Content, % | Recovery, % | Content, % | Recovery, % |
| Magnetic Fraction 1 (MF1) | 15.64 | 28.20 | 94.11 | 0.39 | 17.09 | 0.34 | 43.22 |
| Magnetic Fraction 2 (MF2) | 4.66 | 3.51 | 3.50 | 3.15 | 41.60 | 0.80 | 30.70 |
| Non-Magnetic Fraction (NMF) | 79.70 | 0.14 | 2.40 | 0.18 | 41.31 | 0.04 | 26.08 |
| Feed | 100.00 | 4.69 | 100.00 | 0.35 | 100.00 | 0.12 | 100.00 |

The separation of the iron-bearing silicate minerals from the leaching residue in the magnetic fraction 1 is feasible with a magnetic field intensity of 0.5 T. Over 90% of Fe is recovered in the magnetic fraction 1 with a yield of 15.6%. The XRD pattern of the magnetic fraction is presented in Figure 7.

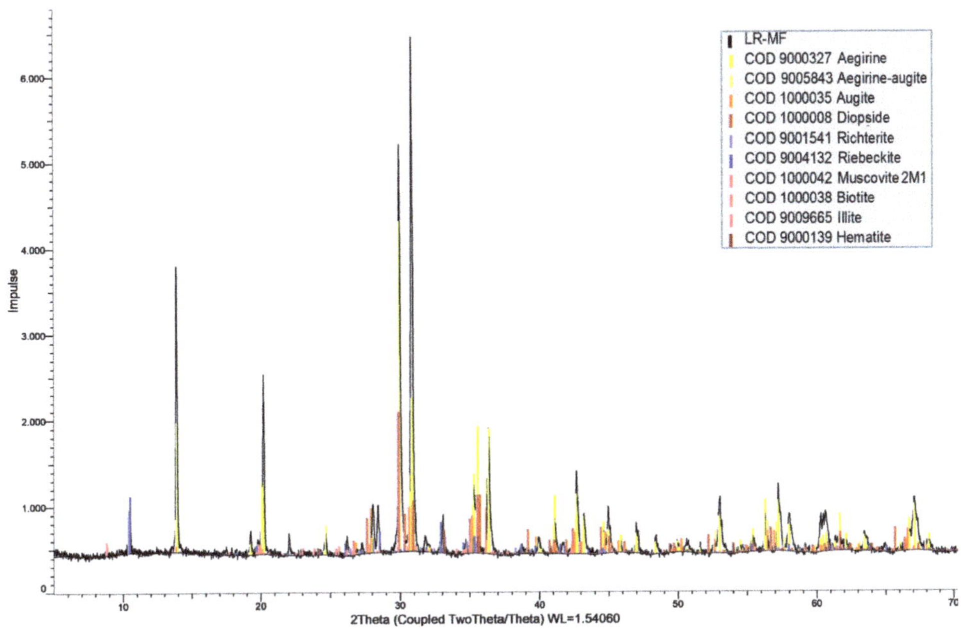


**Figure 7.** XRD pattern of the magnetic fraction.

The major phases belong to aegirine-augite series (89%), alkali amphiboles (riebeckite) and micas. The aegirine particles are elongated and needle-shaped crystals (as shown in Figure 8a).

Thus, it was also confirmed that after hydrometallurgical treatment using a prototype digestion reactor, the mafic minerals preserve their magnetic properties and can be recovered in the magnetic fraction.

The separation of feldspars in the non-magnetic fraction is carried out with a magnetic field intensity of 1.6 T. The main aim of the second magnetic separation stage is to further reduce the iron content in the feldspar product. The middling magnetic fraction (MF2), obtained due to the two-staged magnetic separation, has a $ZrO_2$ content of 3.15% with a mass recovery of 4.6%. The middling fraction is the feed for the resource-efficient recovery of Zr and Hf, which is mentioned by Ma et al. [23]. The amounts of Fe and Ti in the non-magnetic fraction were reduced to 0.14% and 0.04%, respectively. The microscope image of the purified fraction is presented in Figure 8b. According to mineralogical analysis, the non-magnetic fraction consists of feldspars (88%) and residue zeolites (Figure 9).

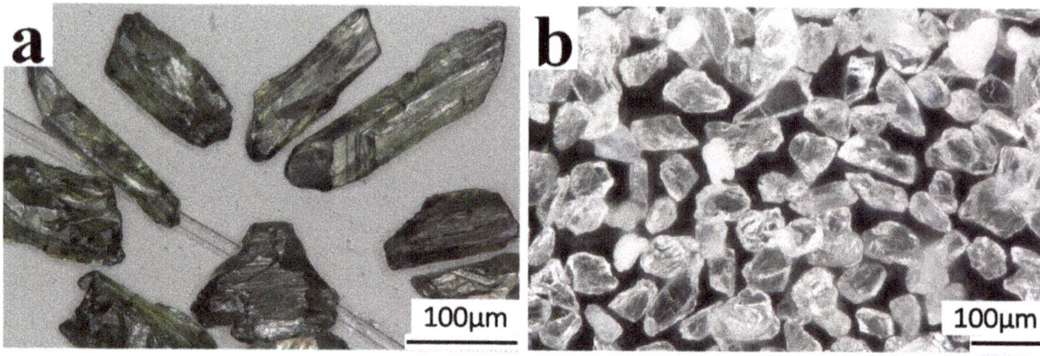


**Figure 8.** Optical microscope images of magnetic (**a**) and non-magnetic fractions (**b**).

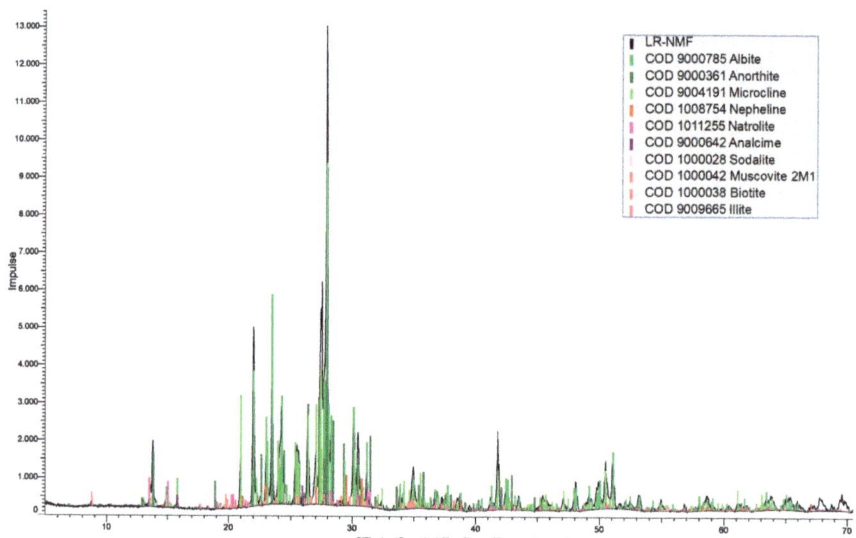


**Figure 9.** XRD pattern of non-magnetic fraction.

Thus, the two-staged magnetic separation is a suitable method to treat the solid leaching residue from eudialyte processing. The chemical composition of the magnetic (MF1) and non-magnetic fractions (NMF) is listed in Table 5.

**Table 5.** Chemical composition of magnetic (aegirine) and non-magnetic fractions (feldspar) (%).

| Products | $SiO_2$ | $Al_2O_3$ | $Na_2O$ | $K_2O$ | CaO | MgO | $Fe_2O_3$ | MnO | $ZrO_2$ | $TiO_2$ | $P_2O_5$ |
|---|---|---|---|---|---|---|---|---|---|---|---|
| Aegirine (MF1) | 48.28 | 1.53 | 8.35 | 0.86 | 1.78 | 2.78 | 28.15 | 0.62 | 0.39 | 0.36 | 0.40 |
| Feldspars (NMF) | 64.10 | 17.21 | 7.10 | 5.11 | 0.55 | 0.16 | 0.14 | 0.01 | 0.18 | 0.04 | 0.15 |

In comparison to the feldspars concentrate, the aegirine concentrate is not presented in the global market as a commodity. The aegirine product composition is comparable with the composition of aegirine concentrate, obtained as a by-product in Khibina deposits from the Kola Peninsula [20]. Currently, the chemical and technical requirements for aegirine concentrates are not specified and must be determined by the potential consumer. Meanwhile, the most critical requirements for the usage of feldspars in glass and ceramic industries are low contents of iron oxide and an alkali content ($K_2O + Na_2O$) as high as possible [17,18]. In comparison to dry magnetic separation, the removal of iron oxide could be improved using wet magnetic separation as well as optimised attrition scrubbing. Equally important is the particle size distribution, because coarse-grained products for glass industry are required [17,18]. The obtained feldspar product could be potentially used for the ceramic production and as filling material. Due to the lower transportation costs, the local consumer might successfully integrate the use the local products. Although the specification for internationally traded concentrates is rigorous, the local consumer might successfully integrate the use of local products due to lower transportation costs.

## 4. Conclusions

A combined treatment starting in a new digestion reactor with filtration, re-pulping, attrition, desliming and a two-step magnetic separation leads to the formation of concentrates of Zr, Nb and Hf from eudialyte ore and the maximal transfer of rare earth elements in the solution, as shown in Figure 10.

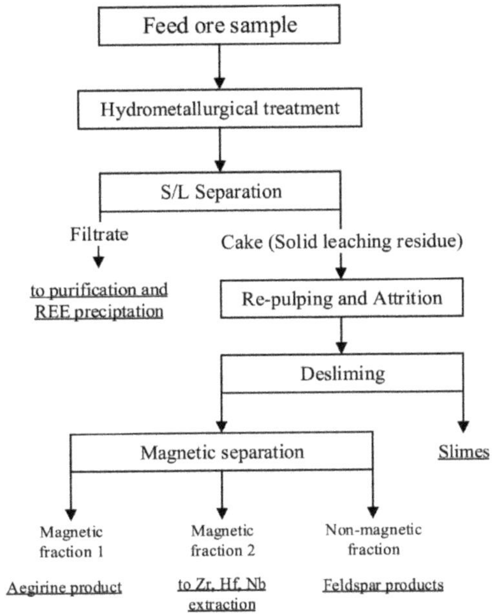


**Figure 10.** Proposed flowsheet for the treatment of a eudialyte ore sample and leaching residue.

During hydrometallurgical treatment with highly concentrated sulphuric acid in a new digestion reactor at room temperature, the leaching efficiency of La, Nd, Ce and Y is higher than 90%. Under the same conditions, silica gel formation is prevented. The solid leaching residue contains the value minerals and can be treated for the production of mineral products. For this purpose, the preliminary deagglomeration of particles of a solid leaching residue using attrition scrubbing is required. Then, the solid leaching residue gets deslimed, and the coarse fraction is treated using magnetic separation. The results of the magnetic separation showed that aegirine- and feldspar-bearing products can be additionally recovered using a two-staged magnetic separation. The WHIMS could be applied to recover aegirine from the slimes. The further application, sales and distribution of marketable mineral products is the challenge for the development of a circular economy. At the other side, the production of rare earth oxides from a solution after purification, precipitation and thermal decomposition is a real option in this case. This developed strategy can be tested for the treatment of bauxite residue in order to decrease treatment costs in the future.

**Author Contributions:** Conceptualization, I.S.; methodology, I.S.; software, I.S.; validation, C.D. and S.S.; formal analysis, S.S. and V.S.C.; investigation, C.D. and I.S.; resources, I.S.; data curation, S.S.; writing—original draft preparation, I.S., S.S. and C.D.; writing—review and editing, I.S., C.D., S.S. and V.S.C.; visualisation, I.S.; supervision, B.F.; project administration, S.S.; funding acquisition, B.F. All authors have read and agreed to the published version of the manuscript.

**Funding:** This research was funded by Programm Zentrales Innovationsprogramm Mittelstand (ZIM)-Kooperation, Funder; Federal Ministry for Economic Afffairs and Climate Action, (No: ZF4098024SU, Acronyme: MAREKO).

**Data Availability Statement:** Data are contained within the article.

**Conflicts of Interest:** The authors declare no conflicts of interest.

## References

1. Dostal, J. Rare Earth Element Deposits of Alkaline Igneous Rocks. *Resources* **2017**, *6*, 34. [CrossRef]
2. Panikorovskii, T.L.; Mikhailova, J.A.; Pakhomovsky, Y.A.; Bazai, A.V.; Aksenov, S.M.; Kalashnikov, A.O.; Krivovichev, S.V. Zr-Rich Eudialyte from the Lovozero Peralkaline Massif, Kola Peninsula, Russia. *Minerals* **2021**, *11*, 982. [CrossRef]
3. Marks, M.A.W.; Markl, G. A global review on agpaitic rocks. *Earth-Sci. Rev.* **2017**, *173*, 229–258. [CrossRef]
4. Vaccarezza, V.; Anderson, C. An Overview of Beneficiation and Hydrometallurgical Techniques on Eudialyte Group Minerals. *Min. Metall. Explor.* **2020**, *37*, 39–50. [CrossRef]
5. Zakharov, V.I.; Maiorov, D.V.; Alishkin, A.R.; Matveev, V.A. Causes of insufficient recovery of zirconium during acidic processing of lovozero eudialyte concentrate. *Russ. J. Non-Ferr. Metals* **2011**, *52*, 423–428. [CrossRef]
6. Lebedev, V.N. Sulfuric Acid Technology for Processing of Eudialyte Concentrate. *Russ. J. Appl. Chem.* **2003**, *76*, 1559–1563. [CrossRef]
7. Lebedev, V.N.; Shchur, T.E.; Maiorov, D.V.; Popova, L.A.; Serkova, R.P. Specific Features of Acid Decomposition of Eudialyte and Certain Rare-Metal Concentrates from Kola Peninsula. *Russ. J. Appl. Chem.* **2003**, *76*, 1191–1196. [CrossRef]
8. Dibrov, I.A.; Chirkst, D.E.; Litvinova, T.E. *Russ. J. Appl. Chem.* **2002**, *75*, 195.
9. Ma, Y.; Stopic, S.; Friedrich, B. Hydrometallurgical Treatment of an Eudialyte Concentrate for Preparation of Rare Earth Carbonate. *Johns. Matthey Technol. Rev.* **2019**, *63*, 2–13. [CrossRef]
10. Stark, T.; Silin, I.; Wotruba, H. Mineral Processing of Eudialyte Ore from Norra Kärr. *J. Sustain. Metall.* **2017**, *3*, 32–38. [CrossRef]
11. Silin, I.; Gürsel, D.; Büchter, C.; Weitkämper, L.; Wotruba, H. Recovery of Catapleiite and Eudialyte from Non-Magnetic Fraction of Eudialyte ore Processing of Norra Kärr Deposit. *Minerals* **2022**, *12*, 19. [CrossRef]
12. Spooren, J.; Binnemans, K.; Björkmalm, J.; Breemersch, K.; Dams, Y.; Folens, K.; González-Moya, M.; Horckmans, L.; Komnitsas, K.; Kurylak, W.; et al. Near-zero-waste processing of low-grade, complex primary ores and secondary raw materials in Europe: Technology development trends. *Resour. Conserv. Recycl.* **2020**, *160*, 104919. [CrossRef]
13. Agrawal, S.; Rayapusi, V.; Dhawan, N. Comparison of microwave and conventional carbothermal reduction of red mud for recovery of iron values. *Miner. Eng.* **2019**, *132*, 202–210. [CrossRef]
14. Liu, Z.; Li, H. Metallurgical Process for valuable elements recovery from red mud—A review. *Hydrometallurgy* **2015**, *155*, 29–43. [CrossRef]
15. Borra, C.R.; Blanpain, B.; Pontikes, Y.; Binnemans, K.; Van Gerven, T. Recovery of rare earths and other valuable metals from bauxite residue (red mud): A review. *J. Sustain. Metall.* **2016**, *2*, 365–386. [CrossRef]
16. Borra, C.R.; Blanpain, B.; Pontikes, Y.; Binnemans, K.; Van Gerven, T. Smelting of bauxite residue (red mud) in view of iron and selective rare earths recovery. *J. Sustain. Metall.* **2016**, *2*, 28–37. [CrossRef]
17. Lorenz, W.; Gwosdz, W. Bewertungskriterien für Industrieminerale. Steine und Erden. Teil 7. Feldspate und andere Flussmittel. Geologisches Jahrbuch, Sonderhefte Reihe H 2004. Heft 10, BGR, Hannover, Germany. Available online: https://www.schweizerbart.de/publications/detail/isbn/9783510959143/Geologisches_Jahrbuch_Reihe_H_Heft_H10 (accessed on 21 January 2024).
18. Militzer, A. *Feldspar Potential in Namibia—Evaluation of Economic Suitability*; BGR: Hannover, Germany, 2020.
19. Popov, I.O.; Mitrofanov, Y.A.; Ustinov, S.M. Feasibility of using aegirine concentrate as a complex flux in copper metallurgy. *Metallurgist* **2012**, *56*, 64–70. [CrossRef]
20. Pleshakov, Y.V.; Fedchenko, V.F.; Shishkin, S.P.; Suvorova, O.V. Aegirite concentrate—Valuable raw material in industry. *Obogashcheniye Rud-Miner. Process. J.* **2005**, *5*, 39–41.
21. Brylyakov, Y.Y. Prospects for all-round utilization of apatite-nepheline ores from the Khibini deposits. *Obogashcheniye Rud-Miner. Process. J.* **2005**, *3*, 28–31.
22. Stopic, S.; Dertmann, C.; Gulgans, U.; Friedrich, B. Neuer modularer Reaktor zum trockenen Aufschließen silikatreicher Erze und Konzentrate und für die Extraktion kritischer Metalle. *Erzmetall-World Metall.* **2019**, *72*, 3.

23. Ma, Y.; Stopic, S.; Gronen, L.; Friedrich, B. Recovery of Zr, Hf, Nb from eudialyte residue by sulfuric acid dry digestion. *Hydrometallurgy* **2018**, *181*, 206–214. [CrossRef]
24. Hedrich, S.; Breuker, A.; Martin, M.; Schippers, A. Rare earth elements (bio)leaching from zircon and eudialyte concentrates. *Hydrometallurgy* **2023**, *219*, 106068. [CrossRef]
25. Judge, W. Azimi Recent progress in impurity removal during rare earth processing: A review. *Hydrometallurgy* **2020**, *196*, 104435. [CrossRef]

*minerals*

MDPI

*Article*

# The Effects of Wind Velocity on the Binding Properties of Ash, Bottom Ash, and Additives: A Wind Tunnel Study

**Sandra Petković Papalazarou [1,*], Jasmina Nešković [1], Stevan Ćorluka [2], Svetlana Polavder [1], Aleksandar Mitrašinović [3,4] and Pavle Stjepanović [1]**

1 Mining Institute, Batajnicki Put 2, 11080 Belgrade, Serbia; jasminanegrojevic76@gmail.com (J.N.); svetlana.polavder@ribeograd.ac.rs (S.P.); pavle.stjepanovic@ribeograd.ac.rs (P.S.)
2 Institute for Materials Testing—IMS, 11000 Belgrade, Serbia; stevan.corluka@institutims.rs
3 The Department of Materials Science and Engineering, University of Toronto, Toronto, ON M5S 3E4, Canada; amitrasinovic@itn.sanu.ac.rs
4 Institute of Technical Sciences of the Serbian Academy of Sciences and Arts, 11000 Belgrade, Serbia
* Correspondence: sandra.petkovic@ribeograd.ac.rs

**Abstract:** The more economically viable and environmentally sustainable approach for treating the by-products of coal combustion from thermal power plants entails their collective disposal as opposed to individual disposal methods. This aligns with pertinent EU directives and domestic regulations, ensuring compliance with established standards while optimizing resource utilization and minimizing environmental impact. This study evaluated the resistance to wind erosion of the binding properties of a mixture (fly ash (FA), bottom ash (BA), and additives) using an indoor wind tunnel under simulated ambient conditions. Investigations of the mutual impact of ash, bottom ash, and additives (CaO and $Ca(OH)_2$) with a certain percentage of water were carried out with eighteen samples. The samples consisted of the water at six addition rates 5, 8, 10, 15, 20, and 25% ($w/w$), and additive at three addition rates (1, 2, and 3% ($w/w$). Based on the obtained results, the optimal ratios of additives (3% ($w/w$)) and water (15% ($w/w$)) were determined. Prior to the wind tunnel experiments, and according to the different addition rates of additives and water, eight samples were prepared with different addition rates of ash. The mass concentrations of suspended particles ($PM_{10}$) and total suspended particles (TSPs) in these samples were measured at three distinct wind velocities: 1 m/s, 3 m/s, and 5 m/s, respectively. The results indicate that the samples containing the optimal content of additives and water demonstrate a maximum increase in $PM_{10}$ emission zero values of no more than 1.9 times. This finding can be considered satisfactory from the standpoint of environmental protection.

**Keywords:** fly ash; bottom ash; additives; wind tunnel; $PM_{10}$; disposal

## 1. Introduction

Coal is a vital energy source that has been developing globally for centuries [1,2]. While coal demand is declining in the United States and may decrease in the European Union this year, global coal consumption still rose by 1.4 percent in 2023, reaching a record 8.5 billion metric tons. The capacity for coal power plants currently under development increased by 16 percent [3]. Because the marginal cost of maintaining coal energy is lower than that of exploiting and using renewable and sustainable energy resources, and current technological advances have lowered the production and operation costs of traditional fossil energy sources while increasing energy efficiency, coal will continue to be consumed for a long time [4]. This trend has also been reinforced by turbulent global political situations. The global energy crisis triggered by the Russia–Ukraine military conflict, which began in February 2022, compelled Europe to revert to a coal-based power generation and energy consumption structure [5].

One of the more important factors is surface coal mines, where coal is stored and used, both in Serbia and worldwide [6]. Countries such as China, Australia, the USA, and others have a large number of surface coal mines [7]. However, due to the characteristics of surface coal mines, a lot of coal dust is produced in the process of storing, transporting, processing, and using coal, which seriously reduces the atmospheric visibility [8]. Apart from coal dust, the biggest problems are FA and BA, which are produced in the process of burning coal in thermal power plants (TPPs). FA is a kind of powdered mineral residue resulting from the combustion of coal which is captured from fly gas, and BA is heavier and harder to carry out with fly gas. Depending on the coal type and source, the chemical composition of BA is mostly similar to FA, mainly composed of silica, alumina, iron, calcium, alkalis with smaller percentages of calcium, magnesium, sulfates, and in some cases also heavy metals, but with a greater content of unburned carbon [9–11]. In recent years, with the continuous development of thermal power, the emission of fly ash (FA) has increased rapidly. Meanwhile, the dust from ash yards has caused serious secondary pollution to the local ecological environment. Therefore, it is necessary to explore surface solidification technologies to prevent FA particles, which contain significant amounts of heavy metals, from dispersing into surrounding cities via wind. This dispersal can lead to severe haze and poses a great threat to the life and health of inhabitants.

The presence of heavy metals in ash can also lead to animal and plant poisoning and DNA damage, threatening the local ecological balance and necessitating the urgent stabilization of ash fields. Various dust control technologies have been employed in ash fields, including engineering methods, water sealing, chemical treatments, and vegetation cover [12]. However, the spray water sealing method, while easy to operate, requires frequent application and often yields unsatisfactory results due to the high porosity and poor water retention of FA [13].

Inhalation of polluted air containing high concentrations of particles or gasses such as $SO_2$ is associated with short-term and long-term health problems, primarily affecting the respiratory and cardiovascular systems [14]. Approximately two million deaths worldwide annually are attributed to air pollution, primarily due to respiratory diseases and cancer [15]. The solidification/stabilization (S/S) method is a widely used treatment for the management and disposal of a broad range of contaminated wastes, particularly those contaminated with substances classified as hazardous. As a result, many different types of hazardous wastes are treated with different binding agents such as lime, cement, gypsum, slag, fly ash, or phosphate cements [16–19].

The disposal of ash, bottom ash, ash-and-bottom ash mixtures, formed at the combustion of coal, peat, and shale, is one of the main problems of the modern solid-fuel thermal power plants in Serbia. The existing system of disposing of combustion products at all landfills of thermal power plants in Serbia is the hydraulic transport of ash, bottom ash, and water mixtures with solid-to-liquid phase ratio S:L—1:10 (low-density hydromixture) and S:L—1:1 (high-density hydromixture). Adding additives to this mixture requires a reduction in the amount of water used for transport, and this significantly influences a reduction in the pollution of air, watercourses, surrounding land, and ecosystems and increases landfill stability [20].

In order to conduct a thorough evaluation of the interconnectedness and impact of FA, BA, and additives on enhancing the binding properties of the mixture, it was essential to identify the fundamental parameters required for the adoption of the most suitable disposal technology. To ascertain the optimal mass fraction of ash, bottom ash, additives, and water within the mixture, it was imperative to replicate landfill conditions as closely as possible in laboratory settings. A specialized wind tunnel, [21–23] has been constructed to replicate wind velocities ranging from 1 m/s to 5 m/s. This wind tunnel includes instrumentation for accurately measuring the concentration of $PM_{10}$ and TSP, alongside capabilities for monitoring relative humidity and temperature. This development represents a novel approach in Serbian research, as the existing literature indicates no prior utilization of wind tunnels for studying solidified material at thermal power plant landfills in Serbia.

The principal objective of this investigation is to ascertain the influence exerted by combustion coal by-products originating from thermal power plants on the ambient air quality within close proximity to such facilities, where concentrations of $PM_{10}$ often exceed prescribed legal air quality thresholds, with heightened prevalence observed during the winter season. This study is motivated by the imperative to discern an efficient technological framework for the responsible management of combustion by-products at all landfill sites with thermal power plants across Serbia.

## 2. Materials and Methods

### 2.1. Materials

The materials used in this research include fly ash and bottom ash procured from Thermal Power Plants in Kostolac, Serbia. As additives, calcium oxide (CaO) and calcium hydroxide ($Ca(OH)_2$) from Kolubara Građevinar, Lazarevac, Serbia, were incorporated into the mixture.

For the preparation of the mixture, a comprehensive characterization of the fly ash and bottom ash samples was conducted. This included particle size analysis, the determination of bulk density in both loose and compacted states, and chemical composition analysis. These analyses were pivotal in selecting the optimal quantities of materials for subsequent testing in the wind tunnel. A detailed description of the methodologies employed in these analyses is provided below.

### 2.2. Methods

#### 2.2.1. Characterization of the Initial FA and BA Samples

The particle size analysis of both fly ash and bottom ash was carried out using standard laboratory sieves. The sieving process was performed manually using Laboratory Test Sieves from Endecotts LTD, London, UK. The measurement of bulk density in both loose and compacted states was conducted in accordance with the Standard EN 1097-3:1998 [24]. The chemical composition of the samples was elucidated through silicate analysis, utilizing the Inductively Coupled Plasma-Optical Emission Spectroscopy (ICP-OES) technique. This analysis was performed using a Varian 710-ES axial ICP-OES spectrometer manufactured by Varian, located in Houten, The Netherlands [25].

#### 2.2.2. Sample Preparation Protocol for Geo-Mechanical Testing

The program was designed to vary three factors: the quantity of water, the quantity and type of additive, and the aging duration post-test specimen's formation. Three distinct groups of test specimens were subsequently created:

- base (initial) test specimens containing only ash and water without additives,
- test specimens with additive (ash + additive + water), and
- test specimens with additive and bottom ash (ash + bottom ash + additive + water).

The authors established the percentage of added water within the range of 5 to 25%, relative to the dry weight of the sample, and the percentage of the additive ranging from 1 to 3%. Specifically, the test specimens were prepared using varying combinations of water and additives:

- Water content (% $w/w$): 5%, 8%, 10%, 15%, 20%, 25%;
- Additive content (% $w/w$): 1, 2, and 3.

Specifically,

Combinations included 5% water with additives at 1%, 2%, and 3%; 8% water with additives at 1%, 2%, and 3%; 10% water with additives at 1%, 2%, and 3%; 15% water with additives at 1%, 2%, and 3%; 20% water with additives at 1%, 2%, and 3%; and 25% water with additives at 1%, 2%, and 3%.

The aging period for the test specimens was set at 1, 7, and 28 days to evaluate their performance over time. All of the aforementioned factors are consistent with the relevant literature. The test specimens were prepared in the laboratory under consistent ambient conditions. The procedure for forming all the test specimens was identical and encompassed the following steps:

- Measurement of the required quantities of water, fly ash, bottom ash, and additives according to the specifications outlined in Table 1.
- Components were mixed and homogenized in a laboratory planetary mixer (manufactured by "Toni Technik") for a duration of 5 min.
- Addition of the specified amount of water to the mixture, followed by further stirring for an additional 5 min.
- After 30 min of aging, the prepared mixture was poured into a steel mold (with dimensions of D × H = 50 × 81 mm) and subjected to a compacting force of 120 $kN/m^2$ using a laboratory hydraulic press (manufactured by W. Feddeler, Essen, Germany).
- Following extrusion from the mold, test specimens were measured, and their dimensions and weight were recorded. They were then appropriately labeled; subsequently, the test specimens were placed for aging under laboratory conditions, with ambient temperature maintained between 20 and 21 °C and relative humidity between 40 and 50%, while ensuring no exposure to direct sunlight.
- The aging period was 1, 7, and 28 days, following which the test specimens were subjected to geo-mechanical tests. Prior to subsequent investigation, the test specimens were measured after aging, and their weight and dimensions were recorded.
- The formed test specimens exhibited a cylindrical shape, with a diameter of 50 mm and a height of 51 ± 1 mm, as illustrated in Figure 1.

**Table 1.** Comparison of wind velocities in wind tunnel and at 10 m height.

| Velocity in Wind Tunnel, m/s | Wind Velocity at 10 m, m/s |
|---|---|
| 1 | 2.82 |
| 3 | 8.45 |
| where: | |
| $z$ | 10 m |
| $z_0$ | 0.0003 m—according to the European Wind Atlas for sand surfaces (smooth) |
| $z_r$ | 0.05 m—the height in the wind tunnel at which the velocity is measured |
| $\alpha$ | $\alpha = \left(\frac{1}{\ln\left(\frac{z_r}{z_0}\right)}\right) = 0.17$ |

**Figure 1.** Geo-mechanical test specimens.

2.2.3. Preparation of New Mixture Samples for Wind Tunnel Testing

Based on the results obtained from the geo-mechanical tests, the most promising combinations of components were identified. Subsequently, mixtures embodying these combinations were prepared for the determination of dust emission rates in a laboratory wind tunnel. The preparation of mixtures followed steps 1–3. After a 30 min aging process, the mixture was evenly layered into metal trays, covered with lids, and compacted using a laboratory hydraulic press with a force of 120 $kN/m^2$. The trays, measuring $250 \times 500 \times 20$ mm (as depicted in Figure 2), were utilized. For each test conducted in the wind tunnel, a set of 6 trays, identical in size and content, was prepared.

**Figure 2.** Test specimens for wind tunnel testing.

The molds containing samples were positioned within a designated section of the wind tunnel, spanning a length of 1.5 m, as depicted in Figure 3. Concurrently, temperature and relative humidity measurements were conductwed within the same section using a thermo-hygrometer (Comet T3510, COMET SYSTEM, Vsetín, Czech Republic).

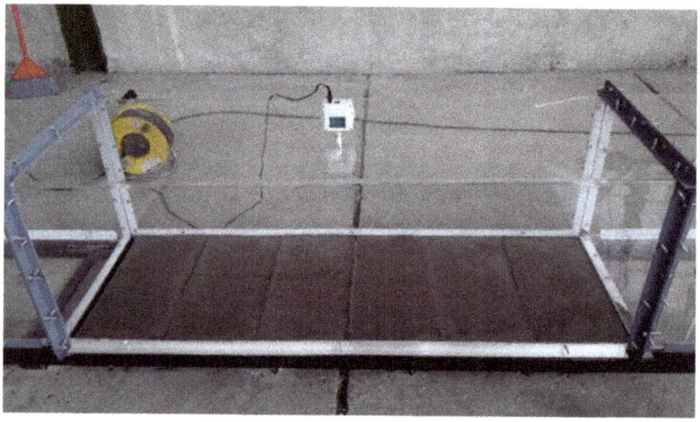

**Figure 3.** Sample tray section within the wind tunnel.

2.2.4. Measurement of Total Suspended Particles (TSPs)

For real-time measurements of the total suspended particles (TSPs) concentration (in $mg/m^3$), a Micro Dust device (Casella Cel-712 PRO) (Casella, Bedford, UK) was employed, utilizing the light scattering method as its measuring principle. Prior to commencing measurements, zero and span values were calibrated. The measurement interval was set to 1 s, with an averaging period of 1 h. The device was positioned 130 cm downstream from the sample section's airflow at a height of 25 cm.

2.2.5. Measurement of Concentration of $PM_{10}$ Particles

The sampling of $PM_{10}$ particles was conducted using a sequential sampler (Sven Leckel KAR8 + MVS6, Sven Leckel, Berlin, Germany), operating at a nominal flow rate of

2.3 $m^3/h$. The measurement results are expressed in micrograms per cubic meter ($\mu g/m^3$), where the air volume is specified at ambient conditions. The weight of the particles was determined by weighing the filter before and after sampling under strictly controlled conditions.

Clean quartz fiber filters (class T293; 47 mm) (Sartorius, Göttingen, Germany) were conditioned in the weighing room at a temperature (20 ± 1) °C and relative humidity 45%–50% before and after sampling for a period of 48 h. The filters were then weighed twice on the analytical scale (Sartorius CPA 225 D-0CE) (Sartorius, Göttingen, Germany) at an interval of 12 h, provided the difference in weight in successive filter weighing did not exceed 40 μg. The filters were then inserted into the sampling head seat, and sampling was carried out at a constant flow of 2.3 $m^3/h$ in a single hour. The sample filter was conditioned for a minimum of 48 h and then weighed on the analytical scale at 24 to 72 h intervals, provided the difference in filter weight did not to exceed 60 μg [26].

### 2.2.6. Calculating the Level of Suspended Particles Emission

The $PM_{10}$ emission level is determined based on the input (min) and the output ($m_{out}$) mass flux through a certain volume (W × D × H). The emission source area (A) is determined as a product of length (L) and width (W), while H represents the height of the wind tunnel. The mass balance for the control volume gives the emission level (E).

$$E = \frac{1}{A}(m_{out} - m_{in}) \tag{1}$$

The mass flux can be determined based on the vertical particle concentration profile (C) and velocity (u) using the following formulas:

$$m_{out} = \int_0^H C_{out} u_{out} V dz \tag{2}$$

$$m_{in} = \int_0^H C_{in} u_{in} V dz \tag{3}$$

which gives the following:

$$E = \frac{1}{L}\int_0^H (C_{out} u_{out} - C_{in} u_{in}) dz \tag{4}$$

If we assume that $u_{out} = u_{in}$, where $u_{out}$ is the velocity of air exiting the control volume and u_in is the velocity of air entering the control volume, we can obtain the equation for the level of particle emission (E, $\mu g/m^2 s$) based on the test area (A) as follows:

$$E = C_{PM10} \times u \times \left(V \cdot \frac{b}{A}\right) \tag{5}$$

where b is the height at which measurement is conducted (0.17 m for $PM_{10}$).

### 2.2.7. Wind Tunnel

The simulation of real conditions is achieved in a straight-line pressure wind tunnel, as depicted in Figure 4, which was designed and constructed by our team. The wind tunnel comprises the following components:

- axial fan with frequency regulator;
- section for air flow conditioning;
- section for directing air flow;
- section with samples;
- section for measurement and sampling.

The tunnel length spans 5.5 m, with a cross-section area of 0.25 square meters. The length-to-height ratio, approximately 11, significantly surpasses the minimum threshold required for forming the boundary layer [27]. Velocity measurements were conducted using a DA4000 anemometer (Pacer Instruments by Miltronics Mfg., Inc, Keene, NH, USA) featuring a measuring range from 0.3 m/s to 35 m/s (±1% accuracy). Positioned along the central axis at a height of 0.25 m, the anemometer is situated 175 cm downstream from the airflow originating from the flow conditioner, as depicted in Figure 4.

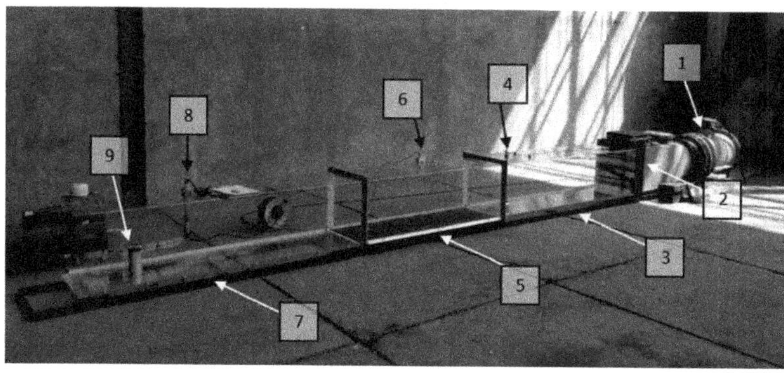


**Figure 4.** Design of the wind tunnel (1—axial fan, 2—flow conditioning section, 3—flow straightening section, 4—wind velocity measurement, 5—sample tray section, 6—temperature and humidity measurement, 7—sampling/measurement section, 8—TSP measurement, 9—$PM_{10}$ measurement).

A grid measurement of flow velocity with the Pitot-Prandtl probe in three axes with five measuring points each (a total of fifteen points) was performed to determine the velocity profile in the cross-section of the tunnel [28].

The logarithmic wind profile has a solid theoretical background compared to the power law wind profile. The standard logarithmic wind profile, used to describe fluid flow over rough surfaces, requires the use of two unknown scaling parameters, the friction velocity, $u_*$, and the aerodynamic roughness length, $z_0$:

$$\frac{U_z}{u_*} = \frac{1}{k} ln\left(\frac{z - z_0}{z_0}\right)$$

where $U_z$ is the wind velocity at height $z$ above the surface, $u_*$ is the friction velocity, $z_0$ is the aerodynamic roughness length of the underlying surface, and $k$ is the von Karman constant, usually assigned a value of approximately 0.4.

The friction velocity ($u_*$) is a scaling velocity of the surface shear stress and is defined by the relationship which depends on the nature of the surface and mean velocity value:

$$u_* = \sqrt{\frac{\tau_0}{\rho}}$$

where $\tau_0$ is the wall shear stress and $\rho$ is the air density.

The power law wind profile describes the mean profile by a simple power function of height as follows:

$$\frac{U_z}{U_r} = \left(\frac{z}{z_r}\right)^{\alpha}$$

where $U_r$ is the known wind velocity at a reference height $z_r$, and $\alpha$ is the power law exponent, which changes with terrain roughness [29,30].

These results demonstrate a correlation between the airflow velocity within the wind tunnel and the corresponding wind velocities at a reference height of 10 m. These data support the effectiveness of our wind tunnel in simulating real-world wind conditions.

The test design used in this study is common in both portable systems [31,32] and laboratory wind tunnels [33], though standards for step magnitude and duration are lacking. This design is efficient for testing numerous surfaces in the field.

Ten samples were prepared to investigate the effect of compression on the stabilization of the deposited mixture of ash and bottom ash. The composition of the tested samples is detailed in Table 2.

**Table 2.** The composition of the tested samples.

| Sample No. | Fly Ash Content % | Water Content % | CaO % | $Ca(OH)_2$ % | Bottom Ash Content % |
|---|---|---|---|---|---|
| U1 | 79 | 15 | - | - | 6 |
| U2 | 76 | 15 | - | 3 | 6 |
| U3 | 76 | 15 | 3 | - | 6 |
| U4 | 74 | 15 | - | 5 | 6 |
| U5 | 84 | 10 | - | - | 6 |
| U6 | 82 | 10 | - | 2 | 6 |
| U7 | 81 | 10 | - | 3 | 6 |
| U8 | 71 | 20 | - | 3 | 6 |

## 3. Results

### 3.1. Particle Size Analysis

The particle size analysis of ash, bottom ash, and additives shows that fly ash predominantly comprises particles within the size class of 150 to 63 µm, commanding a notable share of 61.34%. This observation underscores the significance of this size range in defining the characteristics of fly ash.

Remarkably, hydrated lime ($Ca(OH)_2$) exhibits a marginally augmented prevalence of particles falling within the same size class, reaching a proportion of 80.50%. This modest increase in percentage underscores the similarity in particle size distributions between fly ash and hydrated lime, albeit with a slight variation in prevalence.

Of particular interest is the heightened occurrence of particles within the 150 to 63 µm size class in CaO, which is notably elevated at 93.16%. This pronounced predominance suggests a significant aggregation of particles within this size range, accentuating the distinctive particle size profile of CaO.

The particle size analyses of the investigated samples are meticulously delineated in Figures 5–8.

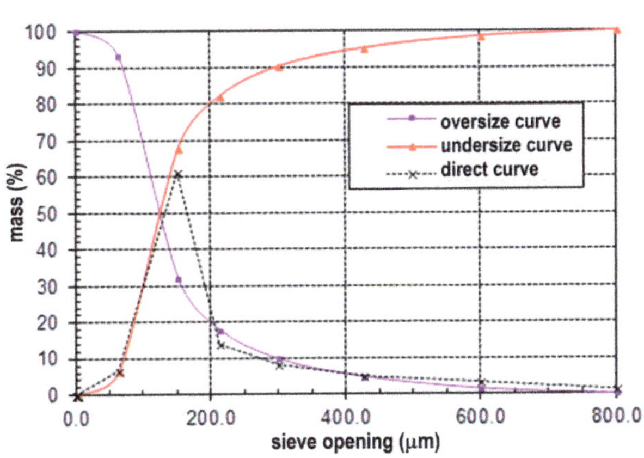


**Figure 5.** Particle size analysis of ash.

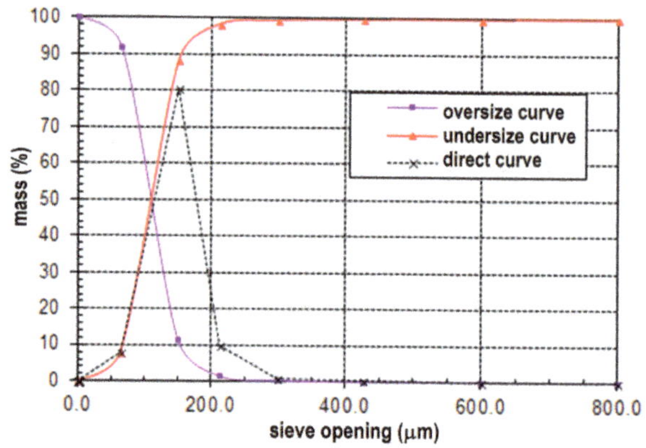


**Figure 6.** Particle size analysis of hydrated lime.

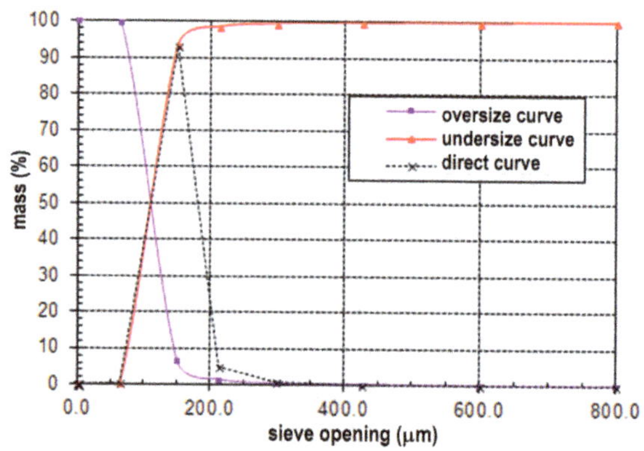


**Figure 7.** Particle size analysis of quick lime.

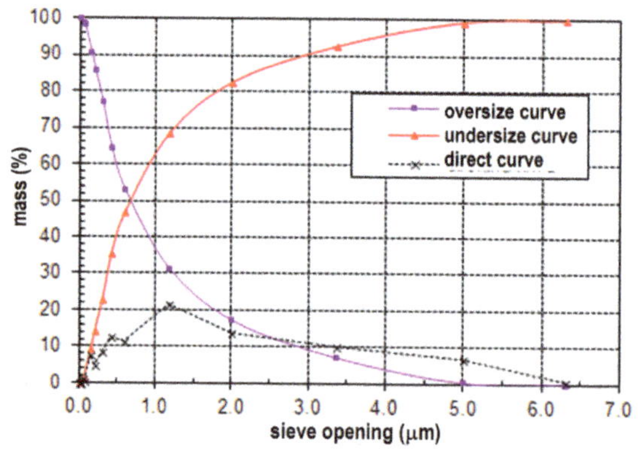


**Figure 8.** Particle size analysis of bottom ash.

Important data for defining certain technical and technological processes, specifically the mean particle diameter (d50) and upper size limit (d95), serve as key indicators for understanding particle size distribution and guiding process optimization efforts. These values can be obtained from Figures 5–8. For the ash sample:

- The mean particle diameter (d50) is determined to be 125 μm.
- The upper size limit (d95) is observed to be 405 μm.

For hydrated lime:

- The mean particle diameter (d50) is measured to be 110 μm.
- The upper size limit (d95) is determined to be 180 μm.

For quick lime:

- The mean particle diameter (d50) is found to be 110 μm.
- The upper size limit (d95) is identified as 160 μm.

For bottom ash:

- The mean particle diameter (d50) is 700 μm.
- The upper size limit (d95) is observed to be 3800 μm.

### *3.2. Results of Chemical Analysis*

Based on the results obtained, the chemical composition of ash is depicted in Figure 9.

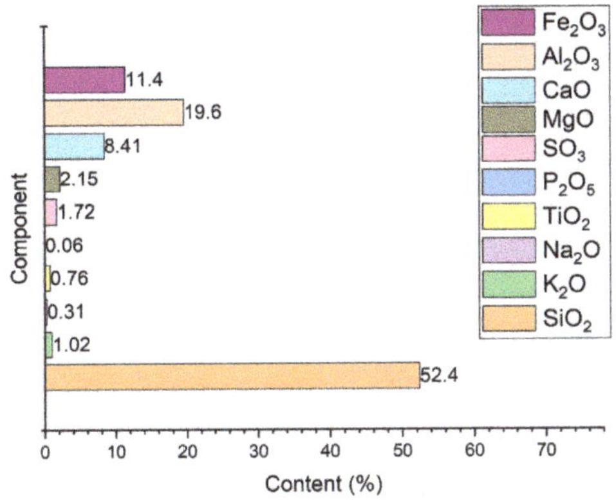


**Figure 9.** Chemical composition analysis of ash.

Based on the presented chemical composition, it is evident that the ash from TPP Kostolac B is classified as silicate ash, with $SiO_2$ being the predominant component at 52.38%. These findings align with ASTM Standard C618 [34], indicating that the ash from TPP Kostolac B falls within the F ash class.

The combined percentage of $SiO_2$, $Al_2O_3$, and $Fe_2O_3$ amounts to 83.34%, surpassing the requirement stipulated by the standard, which mandates it to be more than 70%. The $SO_3$ content is measured at 1.72%, falling comfortably below the threshold set by the standard, which specifies it should be less than 5%. The loss of ignition is calculated to be 2.45%, well within the permissible limit outlined by the standard, which dictates it should be less than 6%.

Bottom ash is generally considered to be an inert material. Its lower chemical activity compared to ash is attributed to the larger size of bottom ash particles and the lower content of $SiO_2$ and $Al_2O_3$, as confirmed by the diagram presented in Figure 10.

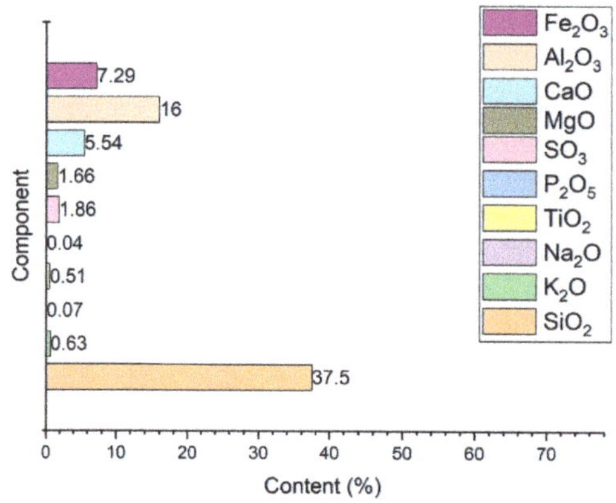


**Figure 10.** Chemical composition analysis of bottom ash.

Based on the obtained results, it can be concluded that both additives, hydrated lime and quick lime, meet the quality standards outlined in the European standard EN 459-1 [35]. This standard specifies criteria regarding chemical composition, purity, reactivity, and other relevant parameters.

In the hydrated lime, the proportion of active $Ca(OH)_2$ is 89.59%, with a total CaO content of 71.39%. The content of MgO is 1.66%, while the combined content of $Fe_2O_3$ and $Al_2O_3$ is 1.7%. In the quick lime, the proportion of active $Ca(OH)_2$ is 91.41%, with a total CaO content of 93.1%. The content of MgO is 0.52%, and the combined content of $Fe_2O_3$ and $Al_2O_3$ is 1.2%.

### 3.3. Results of Bulk Density

The results of the bulk density test in the loose and compacted state ($g/dm^3$) are presented in Table 3.

**Table 3.** Bulk density in loose and compacted state.

| Bulk Density | Loose State | Compacted State |
|---|---|---|
| Sample | ($g/dm^3$) | ($g/dm^3$) |
| Ash | 812.53 | 1032.63 |
| Bottom ash | 467.70 | 533.47 |
| $Ca(OH)_2$ | 497.80 | 671.30 |
| CaO | 642.70 | 849.60 |

The density of the analyzed samples is presented in Table 4.

**Table 4.** Density of analyzed samples.

| Sample | Ash | Bottom Ash | $Ca(OH)_2$ | CaO |
|---|---|---|---|---|
| Density ($g/cm^3$) | 2.1 | 2.1 | 2.4 | 3.6 |

The bulk density of the compacted samples is depicted in Figure 11. The results indicate that, irrespective of the additive amount (1%, 2%, or 3%), the bulk density consistently decreases with aging time, stabilizing after 7 to 10 days. The accompanying diagram illustrates the average percentage weight loss of compacted test specimens relative to the aging time.

The highest weight loss occurs in test specimens composed of mixtures with bottom ash, followed by those without additives, while the test specimens with additives experience the least weight loss with aging.

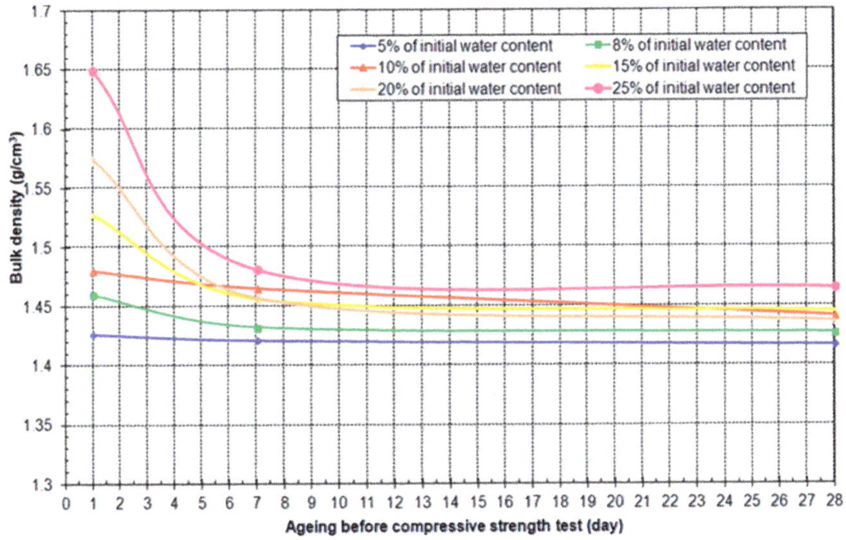


**Figure 11.** Correlation between bulk density and aging time.

*3.4. Results of Geo-Mechanical Testing*

The following characteristic charts will demonstrate the dependence between the one-axial compressive strength and various factors such as water content, additive quantity, and aging time of the compacted mixture.

This chart (Figure 12) clearly indicates that, regardless of the aging time, the most optimal water content is 15%, coupled with 3% of $Ca(OH)_2$ additive. The dependence between one-axial compressive strength after 28 days of aging and water content, as well as hydrated lime content for the ash mixture, is depicted in Figure 13.

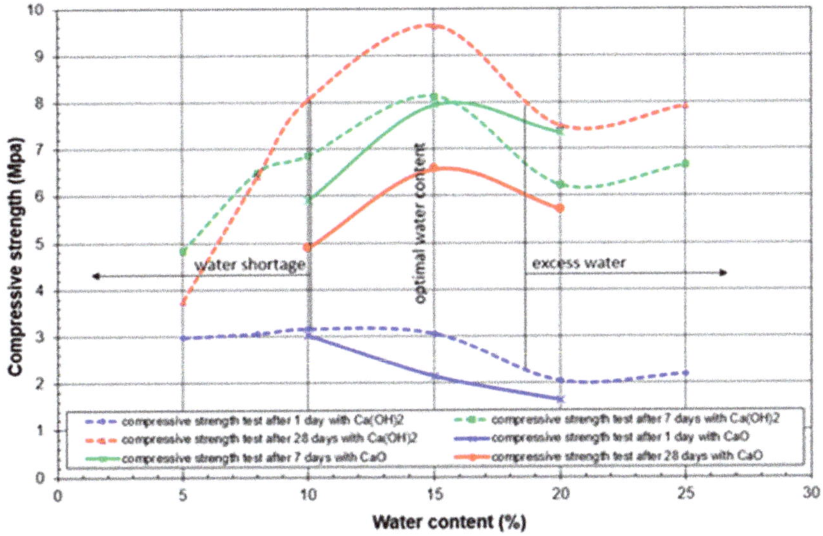


**Figure 12.** Dependence between one-axial compressive strength and water content.

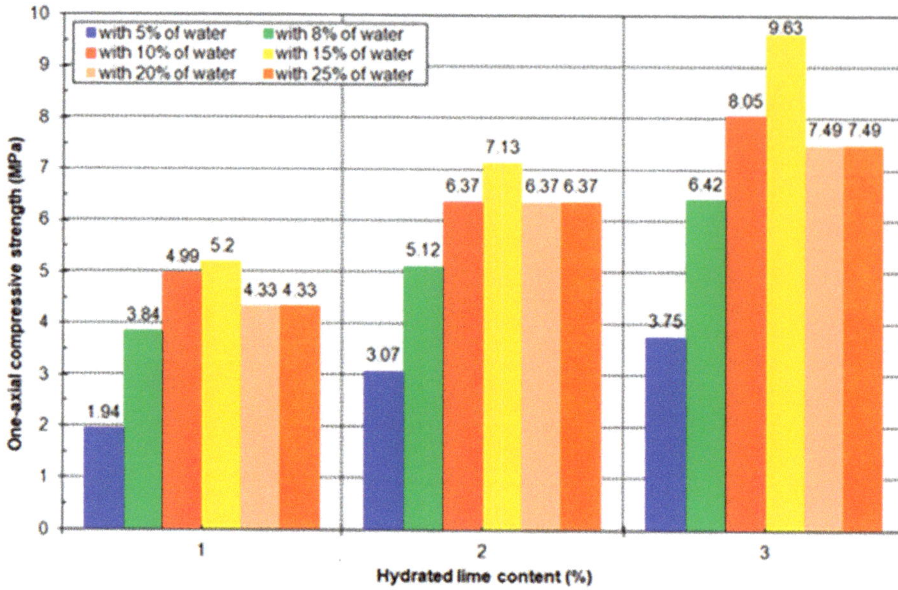


**Figure 13.** Dependence of one-axial compressive strength after 28 days of aging on water and hydrated lime content for the ash mixture.

With an optimal water content of 15%, the compressive strength increases with aging time. This diagram (Figure 14) clearly illustrates that the addition of bottom ash to the mixture significantly reduces the compressive strength.

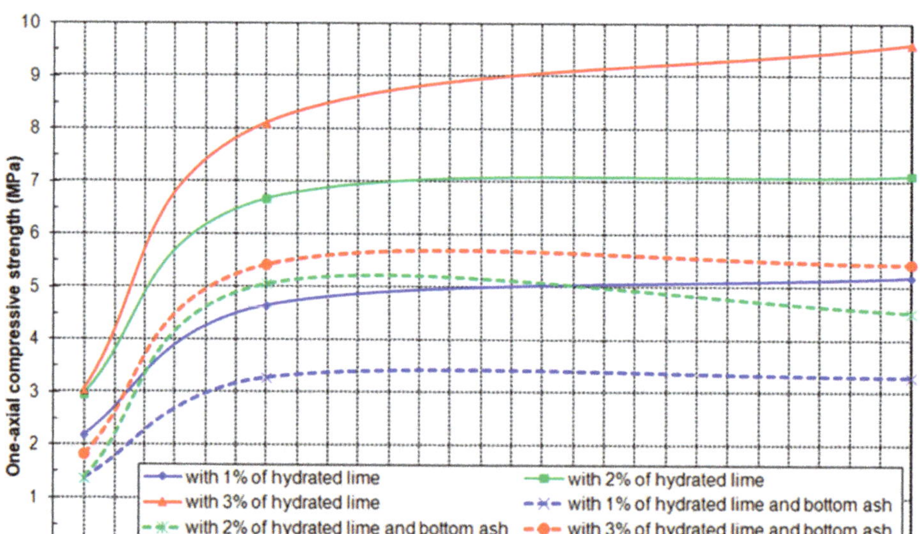


**Figure 14.** Dependence of one-axial compressive strength during aging of ash and bottom ash mixture with 15% water content on hydrated lime content.

The comparison of the previous two diagrams (Figures 15 and 16) clearly shows that the compressive strength decreases with the addition of water beyond the optimal 15%.

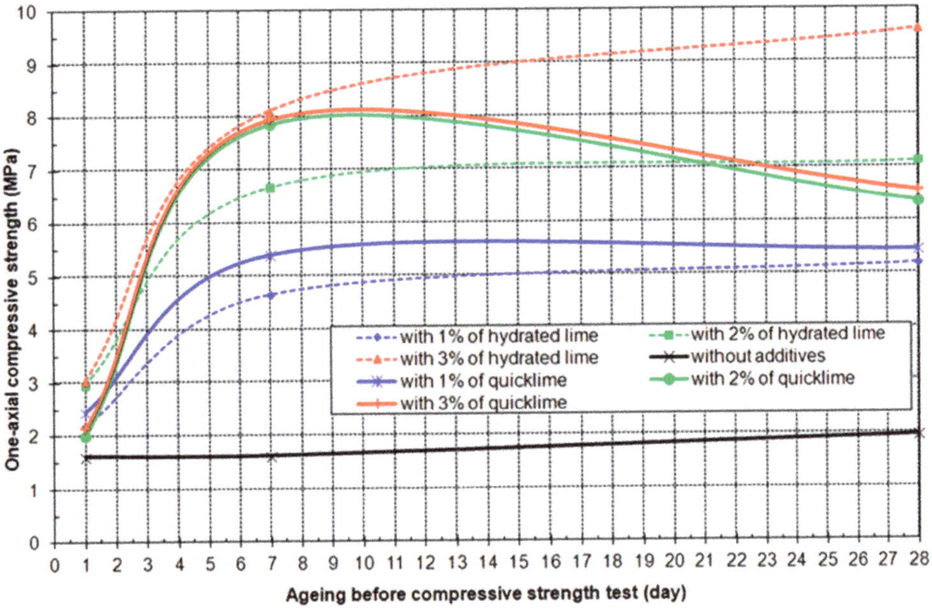


**Figure 15.** Dependence of one-axial compressive strength during aging of ash mixture with 15% of water on percentage of added hydrated/quick lime.

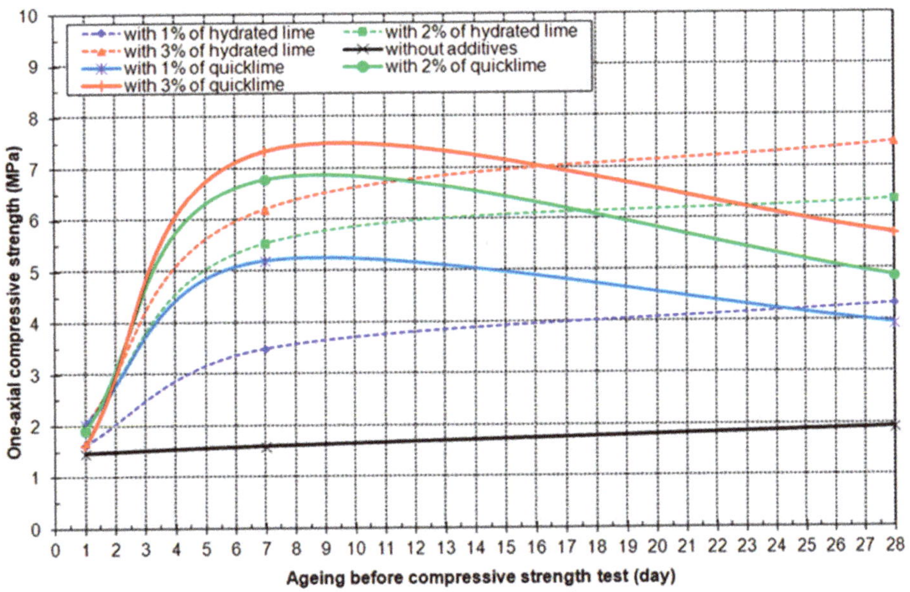


**Figure 16.** Dependence of one-axial compressive strength during aging of ash mixture with 20% of water on percentage of added hydrated/quick lime.

The previous diagram (Figure 17) shows that the highest compressive strength after 28 days is achieved with 15% water and 3% $Ca(OH)_2$. The dependence of the one-axial compressive strength of the ash and bottom ash mixture after 28 days of aging on the percentage of added water and hydrated/quick lime is illustrated in Figure 18.

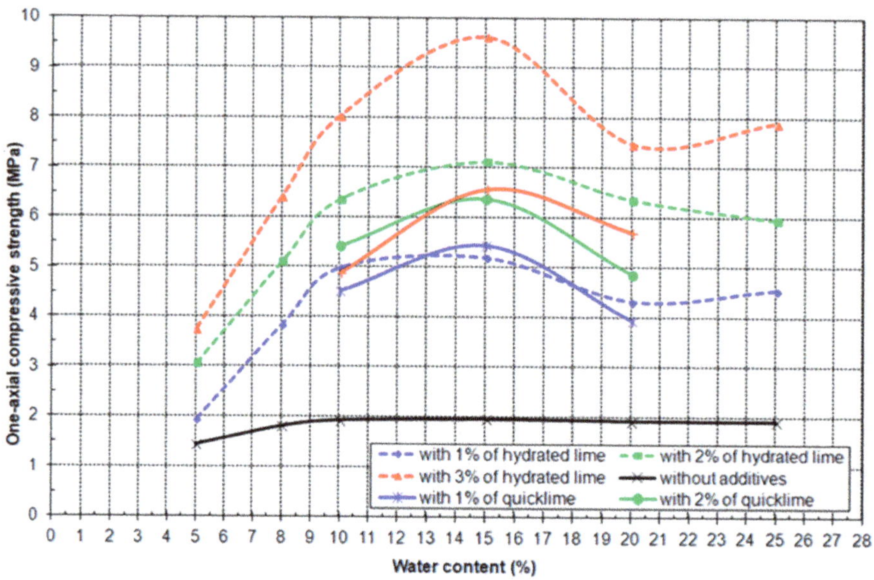


**Figure 17.** Dependence of one-axial compressive strength during aging of ash mixture on the percentage of added water and hydrated/quick lime.

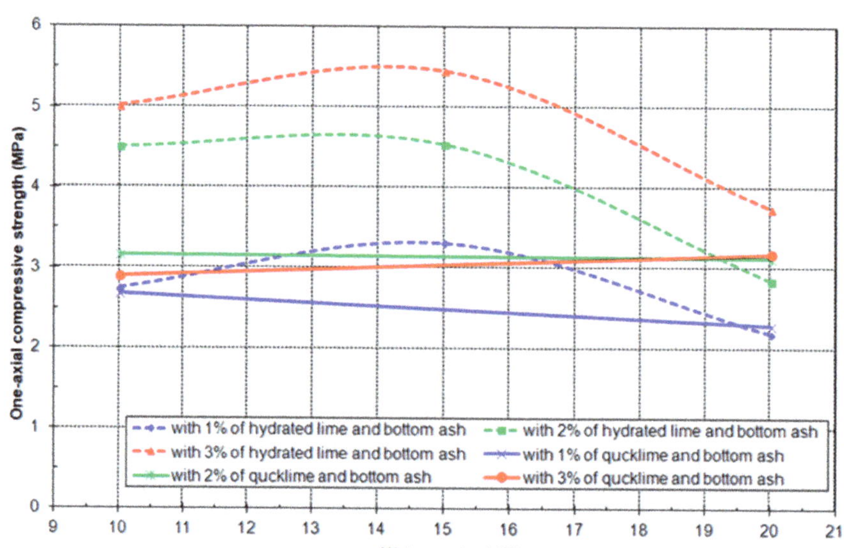


**Figure 18.** Dependence of one-axial compressive strength of ash and bottom ash mixture after 28 days of aging on the percentage of added water and hydrated/quick lime.

The same principle applies when adding bottom ash to the mixture, albeit with a reduction in compressive strength compared to the ash mixture without bottom ash.

With 15% water and 3% $Ca(OH)_2$, the cohesion is most favorable, as clearly visible in the illustrated diagram in Figure 19. However, with the addition of bottom ash, the cohesion decreases, as shown in Figure 20.

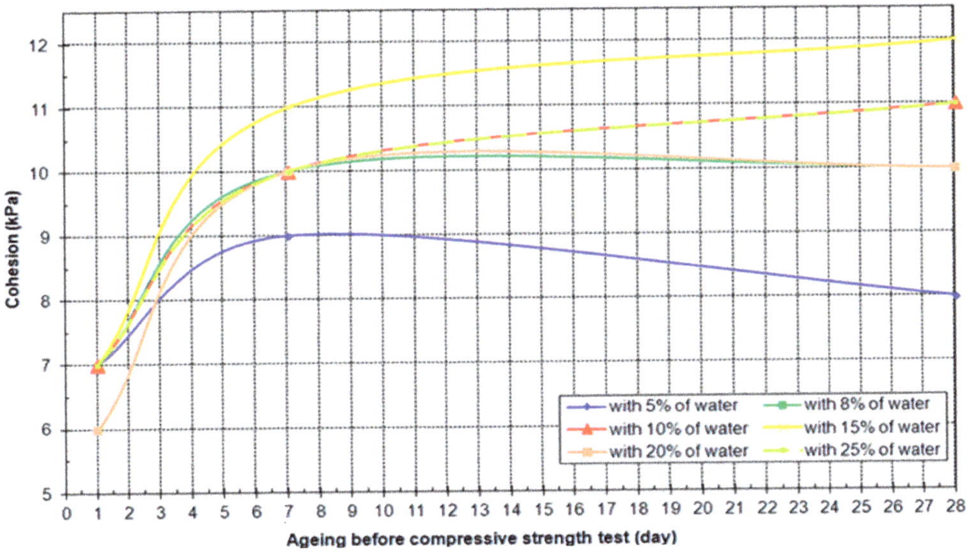


**Figure 19.** The dependence of cohesion of the ash mixture with 3% hydrated lime on the percentage of added water.

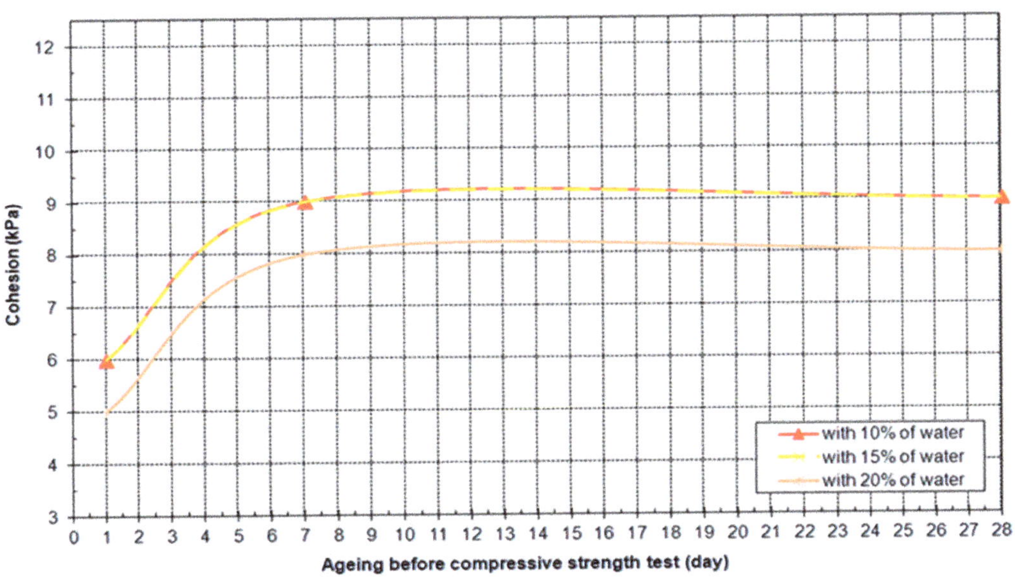


**Figure 20.** Dependence of cohesion of ash and bottom ash mixture with 3% of hydrated lime on aging and percentage of added water.

### 3.5. Results from Wind Tunnel Testing

According to the current domestic regulation [36], the daily mass concentration of $PM_{10}$ suspended particles must not exceed the limit value of 50 μg/m$^3$ more than 35 times in one calendar year. The limit value and the tolerance value for one calendar year are set at 40 μg/m$^3$.

In the absence of an emission limit value, a useful indicator was to compare the obtained values with zero values of the concentration of $PM_{10}$ particles (sample U0) that were measured in the wind tunnel without samples.

The level of emission of $PM_{10}$ particles was calculated as the difference between $PM_{10}$ measured concentrations and $PM_{10}$ zero values. The maximum value of the $EPM_{10}/E0$ ratio was used to estimate the size of the exceeded emission values. The results are shown in Table 5.

**Table 5.** The values of the PM10 emission rates and maximum values of the $EPM_{10}/E0$ ratio.

| Sample No. | Wind Velocity (m/s) | $E_{PM1}$ (μg/m$^2$s) | $E_{PM10}/E_0$ (max) |
|---|---|---|---|
| U0 | 1 | 0.6 | - |
| | 3 | 3.3 | |
| | 5 | 13.8 | |
| U1 | 1 | 2.7 | 4.8 |
| | 3 | 9.9 | |
| | 5 | 13.9 | |
| U2 | 1 | 1.1 | 1.9 |
| | 3 | 3.3 | |
| | 5 | 8.0 | |
| U3 | 1 | 1.1 | 1.9 |
| | 3 | 3.3 | |
| | 5 | 8.4 | |
| U4 | 1 | 0.6 | 20.8 |
| | 3 | 9.9 | |
| | 5 | 286.6 | |
| U5 | 1 | 3.4 | 5.9 |
| | 3 | 13.2 | |
| | 5 | 29.9 | |
| U6 | 1 | 2.7 | 4.8 |
| | 3 | 10.1 | |
| | 5 | 22.0 | |
| U7 | 1 | 2.1 | 3.8 |
| | 3 | 9.9 | |
| | 5 | 18.6 | |
| U8 | 1 | 1.6 | 2.8 |
| | 3 | 8.1 | |
| | 5 | 16.4 | |

As observed from the results presented in Table 4, the $PM_{10}$ emission values escalate with rising flow rates. The highest value of the $EPM_{10}/E0$ ratio is recorded for the U4 sample (without additives and compression), while the lowest value is attributed to samples U2 and U3 (with 15% water and 3% additives $Ca(OH)_2$ and CaO).

The impact of fly ash compression on $PM_{10}$ emission is depicted in Figure 21.

It can be observed that fly ash compression exerts a notable impact on $PM_{10}$ emissions, particularly noticeable at wind velocities exceeding 3 m/s.

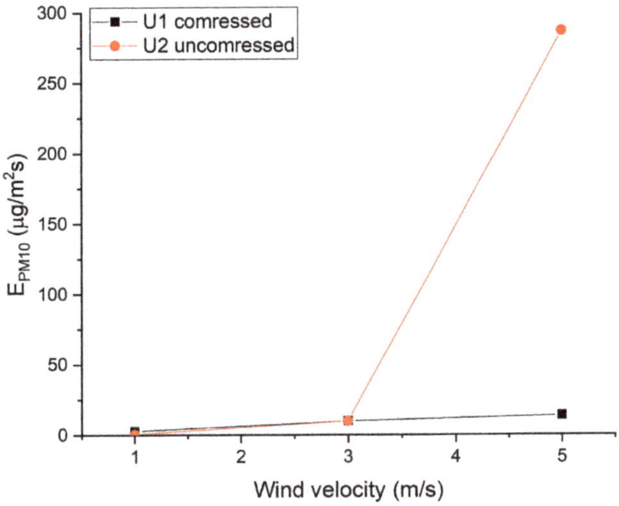


**Figure 21.** PM10 emission rate vs. wind velocity for compressed and uncompressed samples.

The impact of additives on $PM_{10}$ emission is illustrated in the following figure (Figure 22).

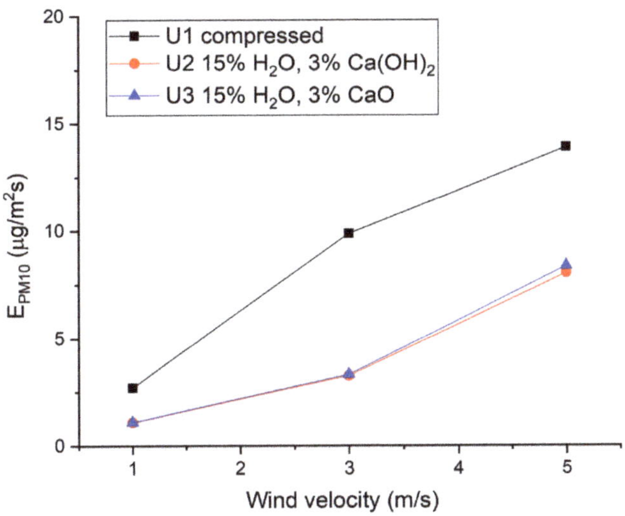


**Figure 22.** $PM_{10}$ emission rate vs. wind velocity for different types of binder.

It is evident that incorporating additives into fly ash leads to a substantial reduction in $PM_{10}$ emissions, and the level of $PM_{10}$ emissions remains consistent when utilizing CaO as an additive compared to $Ca(OH)_2$.

The impact of water content on $PM_{10}$ emission is illustrated in the following figure.

From the results shown in Figure 23, it can be concluded that the lowest $PM_{10}$ emissions were obtained with an additive content of 3%.

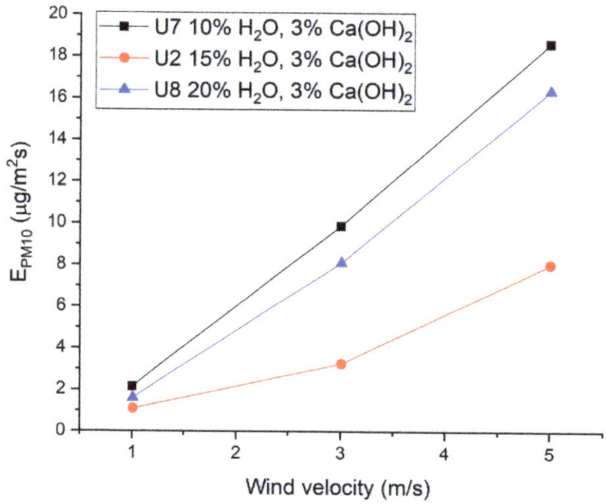


**Figure 23.** $PM_{10}$ emission rate vs. wind velocity for different water contents.

The impact of additives on $PM_{10}$ emission is depicted in the following figure.

As illustrated in Figure 24, the minimum $PM_{10}$ emission level is attained when employing the optimal combination of water content (15%) and additives (3%). Notably, the augmentation of additive content does not yield a discernible impact on $PM_{10}$ emissions, particularly at wind velocities below 3 m/s.

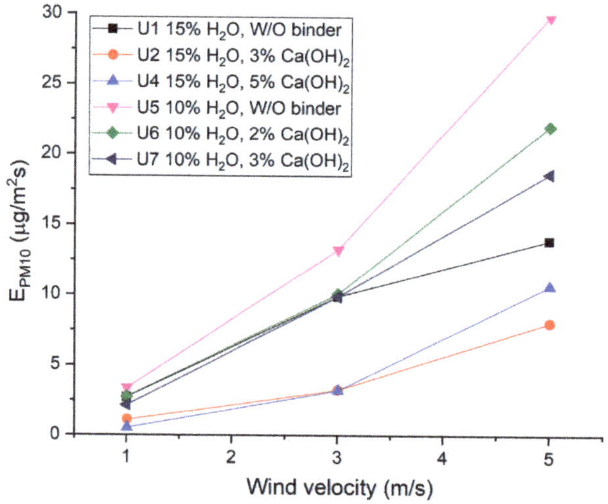


**Figure 24.** $PM_{10}$ emission rate vs. wind velocity for different binder contents.

## 4. Conclusions

The overarching aim of this investigation was to scrutinize the ramifications stemming from the potential dispersion of particles originating from deposited material, impacting both atmospheric and terrestrial environments. This endeavor involved the meticulous evaluation of particle emission levels across eight distinct samples, conducted under varying wind velocities, encompassing velocities of 1, 3, and 5 m per second. The findings suggest that the emission levels observed when incorporating CaO as an additive closely resemble those observed with $Ca(OH)_2$. Moreover, the incremental increase in water content from 15% to 20% (with a 3% additive) does not exert a considerable influence on the $PM_{10}$ emission levels. Furthermore, the augmentation of additive quantity, while maintaining a constant water content, reveals negligible impact on $PM_{10}$ emissions, particularly at wind velocities below 3 m/s.

Drawing upon these findings, it can be deduced that the optimal additive content stands at 3%, while the optimal water content is determined to be 15%. Samples featuring this optimal composition of additives and water demonstrate $PM_{10}$ emission levels that do not surpass zero values by more than 1.9 times. Such outcomes present a satisfactory outcome from an environmental protection standpoint.

**Author Contributions:** Conceptualization, J.N. and S.P.P.; methodology, J.N.; software, P.S.; validation, S.P. and A.M.; formal analysis, S.Ć.; investigation, S.Ć., J.N. and S.P.P.; resources, J.N.; data curation, S.P.; writing—original draft preparation, J.N. and S.P.P.; writing—review and editing, A.M., S.P., J.N. and S.P.P.; visualization, J.N.; supervision S.P.P.; project administration, S.P.P.; funding acquisition, P.S. All authors have read and agreed to the published version of the manuscript.

**Funding:** The authors express their gratitude for the support received from the innovation project of the Ministry of Education, Science, and Technological Development, under project number 391-00-16/2017-16/25, entitled: "Development of technologies for the joint disposal of ash, bottom ash, and gypsum from thermal power plants aimed at improving ecological and economic performances".

**Data Availability Statement:** Data are contained within the article.

**Conflicts of Interest:** The authors declare no conflicts of interest.

## References

1. Chen, Y.; Xu, G.; Huang, J.; Eksteen, J.; Liu, X.; Zhao, Z. Characterization of coal particles wettability in surfactant solution by using four laboratory static tests. *Colloids Surf. A Physicochem. Eng. Asp.* **2019**, *567*, 304–312. [CrossRef]
2. Wang, L.; Liu, Z.; Yang, H.; Li, H.; Yu, M.; He, T.; Luo, Z.; Liu, F. A novel biomass thermoresponsive konjac glucomannan composite gel developed to control the coal spontaneous combustion: Fire prevention and extinguishing properties. *Fuel* **2021**, *306*, 121757. [CrossRef]
3. International Energy Agency. Global Coal Consumption Rises by 1.4% in 2023 to Record 8.5 Billion Metric Tons. 2024. Available online: https://www.iea.org/reports/coal-2023/executive-summary (accessed on 1 June 2024).
4. Chen, J.; Cheng, S.; Song, M.; Wang, J. Interregional differences of coal carbon dioxide emissions in China. *Energy Policy* **2016**, *96*, 1–13. [CrossRef]
5. International Energy Agency. Global Coal Demand Is Set to Return to Its All-Time High in 2022. 2022. Available online: https://www.iea.org/news/global-coal-demand-is-set-to-return-to-its-all-time-high-in-2022 (accessed on 30 May 2024).
6. Han, W.; Zhou, G.; Wang, J.; Jiang, W.; Zhang, Q.; Kong, Y.; Miao, Y. Experimental investigation on combined modification for micro physicochemical characteristics of coal by compound reagents and liquid nitrogen freeze-thaw cycle. *Fuel* **2021**, *292*, 120287. [CrossRef]
7. Wang, P.; Han, H.; Liu, R.; Gao, R.; Wu, G. Effect of outlet diameter on atomization characteristics and dust reduction performance of X-swirl pressure nozzle. *Process Saf. Environ. Prot.* **2020**, *137*, 340–351. [CrossRef]
8. Hall, N.B.; Blackley, D.J.; Halldin, C.N.; Laney, A.S. Current review of pneumoconiosis among US coal miners. *Curr. Environ. Health Rep.* **2019**, *6*, 137–147. [CrossRef] [PubMed]
9. Fidanchevski, E.; Angjusheva, B.; Jovanov, V.; Murtanovski, P.; Vladiceska, L.; Stamatovska Aluloska, N.; Krneta Nikolic, J.; Ipavec, A.; Šter, K.; Mrak, M.; et al. Technical and radiological characterisation of fly ash and bottom ash from thermal power plant. *J. Radioanal. Nucl. Chem.* **2021**, *330*, 685–694. [CrossRef]
10. Singh, N.; Shehnazdeep, A.B. Reviewing the role of coal bottom ash as an alternative of cement. *Constr. Build. Mater.* **2020**, *233*, 117276. [CrossRef]
11. Kim, K.H.; Kabir, E.; Kabir, S. A review on the human health impact of airborne particulate matter. *Environ. Int.* **2015**, *74*, 136–143. [CrossRef]
12. Miao, L.; Wu, L.; Sun, X.; Li, X.; Zhang, J. Method for solidifying desert sands with enzyme-catalysed mineralization. *Land Degrad. Dev.* **2020**, *31*, 1317–1324. [CrossRef]
13. Gadore, V.; Ahmaruzzaman, M. Tailored fly ash materials: A recent progress of their properties and applications for remediation of organic and inorganic contaminants from water. *J. Water Process Eng.* **2021**, *41*, 101910. [CrossRef]
14. Oberdörster, G.; Sharp, Z.; Atudorei, V.; Elder, A.; Gelein, R.; Lunts, A.; Kreyling, W.; Cox, C. Extrapulmonary translocation of ultrafine carbon particles following whole-body inhalation exposure of rats. *J. Toxicol. Environ. Health* **2002**, *65*, 1531–1543. [CrossRef] [PubMed]

15. Mahamaya, M.; Das, S.K.; Reddy, K.R.; Jain, S. Interaction of biopolymer with dispersive geomaterial and its characterization: An eco-friendly approach for erosion control. *J. Clean. Prod.* **2021**, *312*, 127778. [CrossRef]
16. Malviya, R.; Chaudhary, R. Factors affecting hazardous waste solidification/stabilization: A review. *J. Hazard. Mater.* **2006**, *137*, 267–276. [CrossRef] [PubMed]
17. Li, X.D.; Poon, C.S.; Sun, H.; Lo, I.M.C.; Kirk, D.W. Heavy Metal Specifications and Leaching Behaviors in Cement Based Solidified/Stabilized Waste Materials. *J. Hazard. Mater.* **2001**, *82*, 215–230. [CrossRef] [PubMed]
18. Georgescu, M.; Andronescu, C.; Zahanagiu, A. Immobilization of harmful substances in binder matrices. Part I. Immobilization capacity. Influence on binding properties. *Rev. Rom. Mater.* **2006**, *3*, 189–197.
19. Vijan, A.C.; Badanoiu, A.; Voicu, G.; Nicoara, A.I. Phosphate cements based on calcined dolomite: Influence of calcination temperature and silica addition. *Materials* **2021**, *14*, 3838. [CrossRef] [PubMed]
20. Nešković, J.; Stjepanović, P.; Milojković, N.; Lazić, D.; Konc Janković, K. Solidification/stabilization technology of by products (ash) from power plants. In Proceedings of the 8th Balkan Mining Congress, Belgrade, Serbia, 28–30 September 2022; pp. 345–354 ISBN 978-86-82673-21-7. [CrossRef]
21. Xie, S.; Zhang, X.; Pang, Y. Wind Dynamic Characteristics and Wind Tunnel Simulation of Subgrade Sand Hazard in the Shannan Wide Valley of the Sichuan–Tibet Railway. *Int. J. Environ. Res. Public Health* **2022**, *19*, 8341. [CrossRef] [PubMed]
22. Demarco, G.; Martins, L.G.N.; Bodmann, B.E.J.; Puhales, F.S.; Acevedo, O.C.; Wittwer, A.R.; Costa, F.D.; Roberti, D.R.; Loredo-Souza, A.M.; Degrazia, F.C.; et al. Analysis of thermal and roughness effects on the turbulent characteristics of experimentally simulated boundary layers in a wind tunnel. *Int. J. Environ. Res. Public Health* **2022**, *19*, 5134. [CrossRef]
23. Zhao, Z.; Chang, P.; Ghosh, A.; Xu, G.; Liu, Y.; Morla, R. The influence of surfactants' ionicity on the performance of coal dust suppression. *Adv. Powder Technol.* **2024**, *35*, 104295. [CrossRef]
24. *EN 1097-3:1999*; Tests for Mechanical and Physical Properties of Aggregates. Determination of Loose Bulk Density And Voids. iTeh, Inc.: Newark, DE, USA, 1998.
25. Vujić, S.; Radosavljević, M.; Makar, N.; Marijanac, S.; Nešković, J.; Gigov, M.; Žarković, M.; Hafner Ljubenović, T.; Belić, D.; Bužalo, A. Fifty-five years of the Mining Institute. *Bull. Mines* **2015**, *CXII*, 1–27.
26. *BS EN 12341:2023*; Ambient Air. Standard Gravimetric Measurement Method for the Determination of the PM10 or PM2.5 Mass Concentration of Suspended Particulate Matter. European Standards: Plzen, Czech Republic, 2023.
27. Scott Van Pelt, R.; Zobeck, M.T. *Chapter 3, Portable Wind Tunnels for Field Testing of Soils and Natural Surfaces, Wind Tunnel Designs and Their Diverse Engineering Applications*; IntechOpen: London, UK, 2013; Volume 17.
28. *ISO 16911-1:2013*; Stationary Source Emissions Manual and Automatic Determination of Velocity and Volume Flow Rate in Ducts Part 1: Manual Reference Method. International Standards: Geneva, Switzerland, 2013.
29. Tennekes, H. The logarithmic wind profile. *J. Atmos. Sci.* **1973**, *30*, 234–238. [CrossRef]
30. White, B.R. Laboratory simulation of aeolian sand transport and physical modeling of flow around dunes. *Ann. Arid Zone* **1996**, *35*, 187–213.
31. Etyemezian, V.; Gillies, J.A.; Mastin, L.G.; Crawford, A.; Hasson, R.; Van Eaton, A.R.; Nikolich, G. Laboratory experiments of volcanic ash resuspension by wind. *J. Geophys. Res. Atmos.* **2019**, *124*, 9534–9560. [CrossRef]
32. Goossens, D.; Buck, B. Dust dynamics in off-road vehicle trails: Measurements on 16 arid soil types, Nevada, USA. *J. Environ. Manag.* **2009**, *90*, 3458–3469. [CrossRef]
33. Macpherson, T.; Nickling, W.G.; Gillies, J.A.; Etyemezian, V. Dust emissions from undisturbed and disturbed supply-limited desert surfaces. *J. Geophys. Res. Earth Surf.* **2008**, *113*, F02S04. [CrossRef]
34. *ASTM Standard C618-22*; Standard Specification for Coal Fly Ash and Raw or Calcined Natural Pozzolan for Use in Concrete. ASTM International: West Conshohocken, PA, USA, 2023.
35. *European Standard EN 459-1*; Building Lime Definitions, Specifications and Conformity Criteria. iTeh, Inc.: Newark, DE, USA, 2015.
36. *Regulation on Conditions for Monitoring and Air Quality Requirements*; no. 11/10, 75/10 and 63/13; Official Gazette of the Republic of Serbia: Belgrade, Serbia, 2013.

minerals

MDPI

Article

# A Comparative Effect of Sustainable Materials on the Immobilization, Geochemical Fractions, Bioaccumulation, and Translocation of Ni in Smelter- and Mine-Polluted Soils

**Altaf Hussain Lahori [1,2,*], Samreen Riaz Ahmed [3], Ambreen Afzal [4], Monika Mierzwa-Herstek [5,*], Aqsa Umar [6], Maria Taj Muhammad [7], Viola Vambol [8], Irfan Saleem [2], Abrar Ahmad Tanoli [2] and Sergij Vambol [9]**

1 College of Natural Resources and Environment, Northwest A&F University, Yangling 712100, China
2 Department of Environmental Sciences, Sindh Madressatul Islam University, Karachi 74000, Pakistan; irfan8saleem@gmail.com (I.S.); abrargeol786@gmail.com (A.A.T.)
3 Department of English, Sindh Madressatul Islam University, Karachi 74000, Pakistan; samreenriazahmedofficial@yahoo.com
4 National Institute of Maritime Affairs, Bahria University Karachi Campus, Karachi 75260, Pakistan; ambreen.afzal@yahoo.com
5 Department of Agricultural and Environmental Chemistry, University of Agriculture in Krakow, al. Mickiewicza 21, 31-120 Krakow, Poland
6 Department of Computer Science, Sindh Madressatul Islam University, Karachi 74000, Pakistan; aqsa.umar3505@gmail.com
7 Department of Chemistry, Jinnah University for Women, Karachi 74600, Pakistan; mariaataj@gmail.com
8 Department of Applied Ecology and Environmental Sciences, National University «Yuri Kondratyuk Poltava Polytechnic», 36011 Poltava, Ukraine; violavambol@gmail.com
9 Department of Occupational and Environmental Safety, Kharkiv Polytechnic Institute, National Technical University, 61002 Kharkiv, Ukraine; sergvambol@gmail.com
* Correspondence: ahlahori@yahoo.com (A.H.L.); monika6_mierzwa@wp.pl (M.M.-H.)

**Abstract:** The present study holds significant implications, as it aims to investigate the comparative effect of biochar, zeolite, and bentonite minerals on the stabilization of Ni fractions, bioaccumulation, translocation indices, and the reduction in their absorption by pakchoi in smelter- and mine-contaminated soils. The results, which are of great interest, showed that the maximum fresh and dry biomasses of pakchoi were observed up to 28.21 and 18.43% for smelter-polluted soil and 61.96 and 67.90% for mine-contaminated soil amended with zeolite compared to the control. Applying zeolite increased pakchoi chlorophyll SPAD values 1.17-fold in smelter soil and 1.26-fold in mine-polluted soil. The highest Ni immobilization in smelter and mine soil was observed at 76.8 and 85.38% with the application of bentonite, which increased soil pH and CEC. The application of biochar, bentonite, and zeolite reduced the Ni residual, oxidizable, and acid-soluble fractions, but biochar and bentonite increased the reducible fraction of Ni in smelter soil. The highest reduction in Ni in the shoot and root was noted as 82.08 and 68.28% of smelter-polluted soil and 77.25 and 89.61% of mine-polluted soil with bentonite compared to control soil. Overall, it has been concluded that biochar, zeolite, and bentonite can be successfully used to mitigate the Ni concentration in smelter- and mine-polluted soil and reduce uptake by vegetable crops.

**Keywords:** ameliorants; speciation; bioaccumulation induces; soil remediation; plant biomass

## 1. Introduction

Nickel (Ni), rich in the soil system, is a significant threat to the environment. Significantly growing crops and vegetables on Ni-polluted soil can pose a serious health risk to living organisms, which is a great challenging task for researchers [1,2]. The translocation of Ni from the soil pool to the plant body has become a great concern due to human activities, including smelting, metal mining, agrochemicals, electronic industry irrigation, unsafe disposal of spent Ni-Cd batteries, the use of military training and warfare, and fossil fuels consumption [3]. Soil pollution from cadmium, lead, arsenic, mercury, nickel, and copper has been widely reported in the literature. It has a direct or indirect negative impact on humans and ecosystems [4,5]. Nickel has the potential to be toxic, polluting soil and water, posing a threat to food and water security, and hindering sustainable development globally [6]. According to geochemicals, Ni has a +2 valence and is frequently available in most soils, while the highest Ni was found to be in basic rocks and contain high concentrations of humus [7].

Ni is an important micronutrient for crop yield and physiological functions at low levels. Still, its high level may cause negative impacts, including the induction of leaf necrosis, wilting, and chlorosis, and the disruption of photosynthesis, crop production, and soil properties [8,9]. It has an antagonistic impact on plant nutrients viz., Zn, Fe, Ca, Cu, Mn, and Mg [10]. Furthermore, Rattan et al. [11] stated that Ni plays an important role in plants and micronutrients, and its availability at <0.1 mg $kg^{-1}$ may cause plant deficiency. As a result, there is a decreased crop growth, where a higher Ni concentration may show toxicity in mature plant leaves [12]. The occurrence of a high level of Ni over standard values in soil (35 mg $kg^{-1}$) may cause poisonousness to plants, animals, water bodies, microbes, and living organisms [13,14]. The mobility of Ni in the soil pool and its translocation by plants depends on pH, organic matter, soil temperature, microbes, and soil type. With the growing hazard of polluted soil to living organisms, it has become more and more significant to detect and improve soil remediation tools [15]. Nowadays, the perspective courtesy of scholars is to remediate the Ni-polluted soils with the application of feasible and environmentally friendly technologies [16,17].

Several methods are often applied to mitigate the toxicity of pollutants existing in the soils for the safe cultivation of crops and healthy food [18,19]. The in situ approaches comprise stabilization/immobilization, phytoremediation, soil flushing, and microbial remediation [20–22]. Ex situ methods comprise field irrigation, soil washing, and bioreactors [23,24]. Compared with these methods, the soil stabilization method for the fixation of heavy metals, including Ni, has been extensively applied for the restoration of polluted soils because this method has fast results, is easily available, is economically feasible for farming communities, and has a green approach for long-term restoration and safe crop production [25,26]. The recycling of organic waste material into biochar may pay more attention over one decade to the restoration of polluted soils for growing safe crops [27]. The literature reported that several ameliorates have been suggested for the reclamation of polluted soils, including biochar [28] and clay minerals, including bentonite [29,30] and zeolite [31]. Additionally, clay minerals have great potential to stabilize pollutants in soil, soil nutrients, and plant growth and development [32]. Biochar is a solid, black, and carbon-rich material manufactured from the organic residue by the pyrolysis process that can be easily transported and is cost effective compared with fertilizer, and it lasts longer on incorporation than soil [33,34].

The main mechanism of biochar and minerals for the fixation of pollutants in soil systems comprises alkalization, improved ion exchange capacity, and enlarged physical (ad)sorption and precipitation [35,36]. Mihandoost and Rafati [12] assessed the impact of biochar and vermicompost on the phytoavailability of Ni in soil and its absorption by cherry tomato; as a result, Ni absorption in cherry tomato was increased with the application of biochar and vermicompost. Nevertheless, few studies stated that the opposing results, where biochar did not affect or reduce Ni fixation in contaminated environments [37,38]. In the former study, Ali et al. [8] examined the influence of rice straw-made biochar and zeolite on reducing Ni mobility in soil under incubation conditions; as a result, they observed a reduction in extractable Ni in acidic soil with the application of rice straw-made biochar and zeolite at 2%. Shahbaz et al. [39] applied miscanthus feedstock (Miscanthus nepalensis)-made biochar and zeolite in Ni-polluted soil; as a result, Ni was immobilized in soil, and a reduction of the translocation by wheat was observed with the application amendments. Feng et al. [40] assessed the potential of zeolite and biochar in the soil; as a result, they found that pakchoi growth (*Brassica chinensis* L.) was increased with the application of zeolite and biochar. Khan et al. [41] revealed that the translocation and bioaccumulation of heavy metals in soil and agricultural crops did not affect the soil environment and food safety but also had a negative impact on living organisms.

However, there is limited evidence reported on the comparative impacts of tobacco straw waste-made biochar, bentonite, and zeolite regarding the bioavailability of Ni in smelter- and mine-polluted soils. In this study, tobacco straw-made biochar at 1% was used to immobilize the Ni in smelter- and mine-polluted soil and reduce the uptake by pakchoi as a test plant for the first time. It has been observed that the pakchoi crop has broad leaves that have the potential to uptake Ni from polluted soil. Therefore, this study focuses on investigating the comparative individual effects of tobacco straw-made biochar at 1% bentonite, 1% zeolite, and 1% stabilization, bioaccumulation, translocation indices, and geochemical fractions of Ni in smelter- and mine-contaminated soils, and pakchoi growth under in situ condition has not been well studied to date. This work covers a paradigm to an environmentally friendly active method for biochar, bentonite, and zeolite application to immobilize Ni in smelter- and mine-contaminated soils, diminishing their absorption by pakchoi.

## 2. Materials and Methods

### 2.1. Soil Sampling and Amendment Collection

The investigated smelter and mine soils were gathered from the Tongguan and Feng counties of China at the 0–20 cm surface layer. The polyethylene bags were used to transport the bulk samples of contaminated soil to the laboratory. The contaminated soil samples were thoroughly dried for 4 days at room temperature. For the chemical analysis, the dry soil was pulverized and put through a 2 mm sieve. The dried soil was crushed and passed via a 2 mm sieve prior to testing the soil physico-chemical analysis. Furthermore, a 0.25 mm mesh size made up of nylon was used to test the Ni in soil. The prepared tobacco straw-made biochar pyrolysis at (<400 °C) was purchased from Yixin Biological Energy Science and Technology Development Co., Ltd., from Yangliang, China, while Ca-bentonite ($Ca(Al_2Si_4O10)(OH)_2{\cdot}nH_2O$) was purchased from Weifang Huawei Bentonite Group Co., Ltd., Weifang, China and zeolite artificial ($Na_2O.Al_2O_3$. $YH_2O$) was obtained from Zhejiang Shenshi Mining Industry Group Co., Ltd., Lishui, China.

### 2.2. Pot Experiment Set Up

In a pot culture study, 1 kg of air-dried sieved (<2 mm) soil was placed in each pot with a 25 cm height and 20 cm diameter. The application of tobacco biochar 1%, Ca-bentonite 1%, and zeolite 1% dosages were used for stabilizing Ni in smelter- and mine-polluted soils, except for the control treatment [8]. The biochar, bentonite, and zeolite were ground and screened through a 2 mm sieve for soil application. All the immobilizing ameliorants were mixed in both soils, and distilled water was added, aiming to moisten 65% field capacity, and they were incubated for 30 days at room temperature [42]. All three 3 replicated pots were covered with a polyethylene sheet (comprising 10 pinholes) to allow the oxidation process and chemical reaction and to minimize the water evaporation. After the incubation period, the healthy 10 seeds of pakchoi were grown in the individual pot. After germination for 10 days, pakchoi plants were thinned and then kept, and there were only five plants in each pot. The moisture content was adjusted to 80% with distilled water for 2 weeks after germination. No chemical fertilization was amended in the studied pots. After a germination of 65%, distilled water was added to each pot, and the loss waste was adjusted on a daily basis. The plant growth experiment is indicated in Figure 1.

**Figure 1.** Plant growth experiment with the application of biochar, bentonite, and zeolite.

### 2.3. Soils and Amendments Analysis

Soil samples were collected from origin soil after the harvesting of pakchoi in the treated pots for analysis. Approximately 200 g of soil samples were taken from the rhizosphere after the treatment of polluted soil with the application of amendments. The pipette method was used to assess the soil texture of smelter- and mine-polluted soils [43]. To test the soil, pH and EC 1:5 $v/w$ was used in both soils, 1:2 $v/w$ was used for bentonite and zeolite, and (1:10 $v/w$) was used for biochar. Additionally, to determine the dissolved organic carbon, a 1:10 $v/w$ soil and ultra-pure water ratio was used for soils and materials using an automated TOC analyzer Shimadzu TOC-V model [44]. The cation exchange capacity of soils and amendments was examined at pH 7.0 using the ammonium acetate $NH_4CH_3CO_2$ protocol [45]. Humic acid and fulvic acid in both soils were analyzed by Donisa et al. [46]. The surface morphology of biochar, bentonite, and zeolite was examined using scanning electron microscopy (SEM) (TESCAN-LUM), and Fourier-transform infrared (FTIR) (WQF-510A) spectroscopy was used to assess the surface functional groups [47]. The diethylenetriamine pentaacetic acid (DTPA) single extraction method was used to assess the Ni solubility in smelter and mine soils and immobilizing materials by AAS (model: Agilent AA-240FS) followed by the standard Aqua Regia ($HCl–HNO_3–HClO_4$) digestion technique [48]. BCR methods were used to assess the speciation (residual, oxidizable, reducible, and acid soluble) of Ni in both soils after harvesting of the pakchoi plant according to Rauret et al. [49]. Soil and amendment basic properties are indicated in Table 1.

**Table 1.** Soil and amendment chemical properties.

| Parameters | Smelter Soil | Mine Soil | Biochar | Bentonite | Zeolite |
|---|---|---|---|---|---|
| pH | 8.1 ± 0.1 | 8.4 ± 0.3 | 9.6 ± 0.2 | 8.4 ± 0.1 | 7.9 ± 0.4 |
| EC dS/cm | 0.27 ± 0.01 | 0.56 ± 0.2 | 0.55 ± 0.3 | 0.21 ± 0.2 | 0.17 ± 0.1 |
| Sand % | 52.41 | 76.35 | - | - | - |
| Silt % | 46.10 | 23.17 | - | - | - |
| Clay % | 1.49 | 0.48 | - | - | - |
| Textural class | Sandy loam | Loamy sand | - | - | - |
| OM % | 0.26 ± 0.02 | 0.29 ± 0.01 | - | - | - |
| CEC (cmol/kg) | 70.0 ± 0.4 | 53.2 ± 0.2 | 83.05 ± 0.1 | 88.6 ± 0.2 | 91.1 ± 0.3 |

**Table 1.** *Cont.*

| Parameters | Smelter Soil | Mine Soil | Biochar | Bentonite | Zeolite |
|---|---|---|---|---|---|
| DOC mg/kg | 15.1 ± 0.3 | 16.4 ± 0.1 | 37.98 ± 0.4 | 0.01 ± 0.002 | 0.09 ± 0.01 |
| HA % | 0.21 ± 0.1 | 0.26 ± 0.04 | 0.66 ± 0.02 | - | - |
| FA% | 0.17 ± 0.2 | 0.22 ± 0.02 | 0.32 ± 0.01 | - | - |
| Total Ni mg/kg | 92.1 ± 0.2 | 123.5 ± 0.1 | 0.02 ± 0.003 | - | - |
| DTPA-Ni mg/kg | 84 ± 0.3 | 102 ± 0.2 | 0.03 ± 0.01 | 0.10 ± 0.04 | 0.06 ± 0.2 |

Note: pH = soil pH, EC = electrical conductivity, OM = organic matter, CEC = cation exchange capacity, DOC = dissolved organic carbon, HA = humic acid, FA = fulvic acid, Ni = nickel, and DTPA = diethylenetriaminepentaacetic acid.

### 2.4. Plant Analysis

The pakchoi plants were harvested after 45 days of germination. All the pakchoi plant roots were separated from the soil and systematically washed with distilled water. The plant shoot and root biomasses were oven-dried (65 °C) to a constant weight, weighted, and ground with an electric machine and stored in polyethylene bags for chemical analyses. Approximately 0.5 g of dried root and shoot biomasses was used to examine the total Ni concentration in pakchoi root and shoot using concentrated $HNO_3$ (9 mL) and $H_2O_2$ through the microwave-assisted acid digestion method [50]. The SPAD chlorophyll content was noted before harvesting.

### 2.5. Bioaccumulation and Translocation Indices

In the present work, the bioaccumulation and translocation indices were calculated following [51,52], and they are expressed as follows:

BAF = (concentration of Ni in pakchoi root/concentration of Ni in soil).

TF = (concentration of Ni in pakchoi shoot/concentration of Ni in pakchoi root).

### 2.6. Statistical Analysis and Quality Control

The obtained data were statistically analyzed by Statistix 8.1 version. All the graphs were made using Origin-Pro 8.5. The heat map association among the studied parameters was performed using heat map illustrator software (Heml 1.0.3.3). The principal component analysis (PCA) was carried out using Minitab version 17. Certified reference materials, such as GBW07457 (GSS-28 for soil) and GBW07603 (GSV-2 for pakchoi) were used as per the Chinese soil and plant quality control standards. The recovery ratio of Ni in soil and plants was 93%–101% and 94%–102%.

## 3. Results and Discussion

Table 1 reveals that the pH of smelter and mine soils, bentonite and zeolite, were basic in nature, whereas the zeolite pH was slightly alkaline in nature. The soils and additives were non-saline in condition. Both soils were low in OM content due to arable conditions. The smelter soil was sandy loam in nature and the mine soil was loamy sand according to textural class. The soils and additives had exposed medium range in CEC content. The Ni concentration in smelter soil was 1.4 folds and the mine soil was 1.7 folds according to a scale of 60 mg/kg > 7.5 soil pH as per agricultural soil (Environmental Quality Standards, China, GB15618-1995) [53]. SEM spectra present different structures of the materials, in which biochar has more dispersive particles, bentonite consists of significant stratiform particles, and zeolite exhibits the typical mineral structure profile in Figure 2.

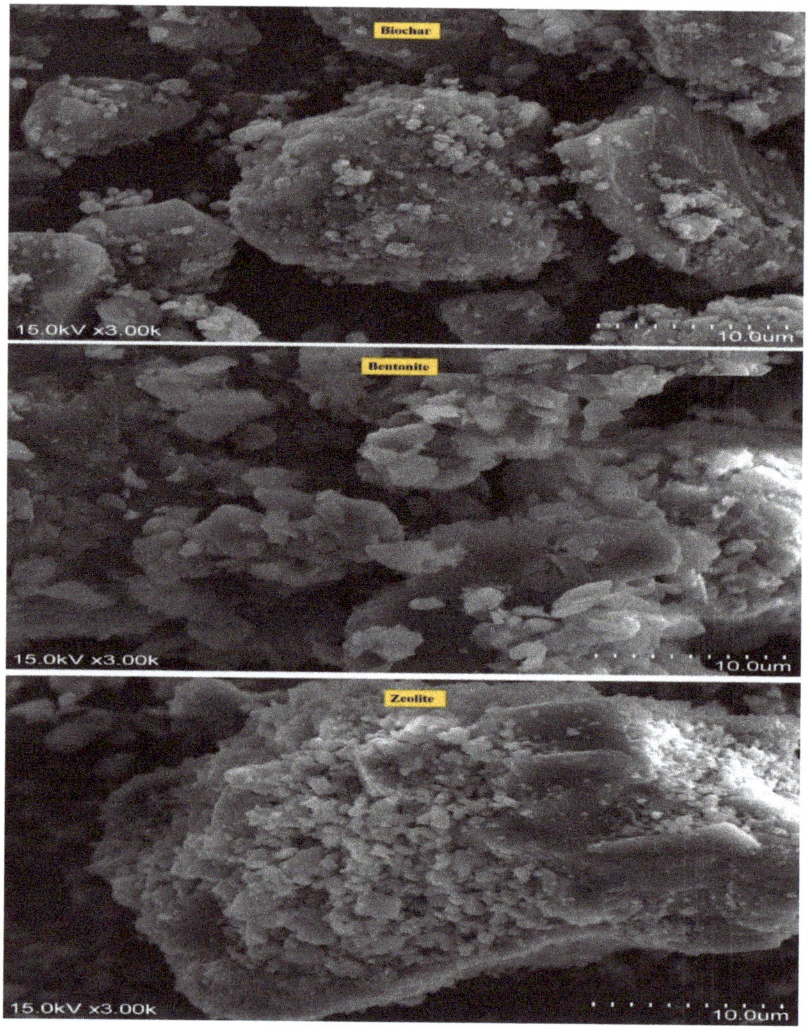


**Figure 2.** Surface electron microscopy (SEM) of biochar, bentonite, and zeolite.

FTIR analysis among the three materials, biochar, bentonite, and zeolite, presents a significant variation in passivation behavior on Ni in smelter- and mine-polluted soil. The strong peaks at 2954.81 and 1426.81 cm$^{-1}$ confirm the formation of a C-H bond in biochar. By contrast, bentonite has a significant characteristic peak of O-H at 3626.23 and 2422.65 cm$^{-1}$, indicating the presence of oxide structures. The strong peak at 1031.61 cm$^{-1}$ reflects the existence of a C-N bond in bentonite, which is probably derived from the organic ammonium complex. Furthermore, many characteristic peaks presenting an O-H bond (at 3625.47, 3422.50, and 2518.05 cm$^{-1}$), C-H bond (at 2979.38, 2873.19, 1431.41, and 1029.42 cm$^{-1}$), O-H bond (at 2518.05 cm$^{-1}$), and C=O bond (1799.90 cm$^{-1}$) are detected in zeolite, reflecting that the adsorption process mainly explains the passivation effect after adding zeolite (Figure 3), according to Shahkolaie et al. [47].

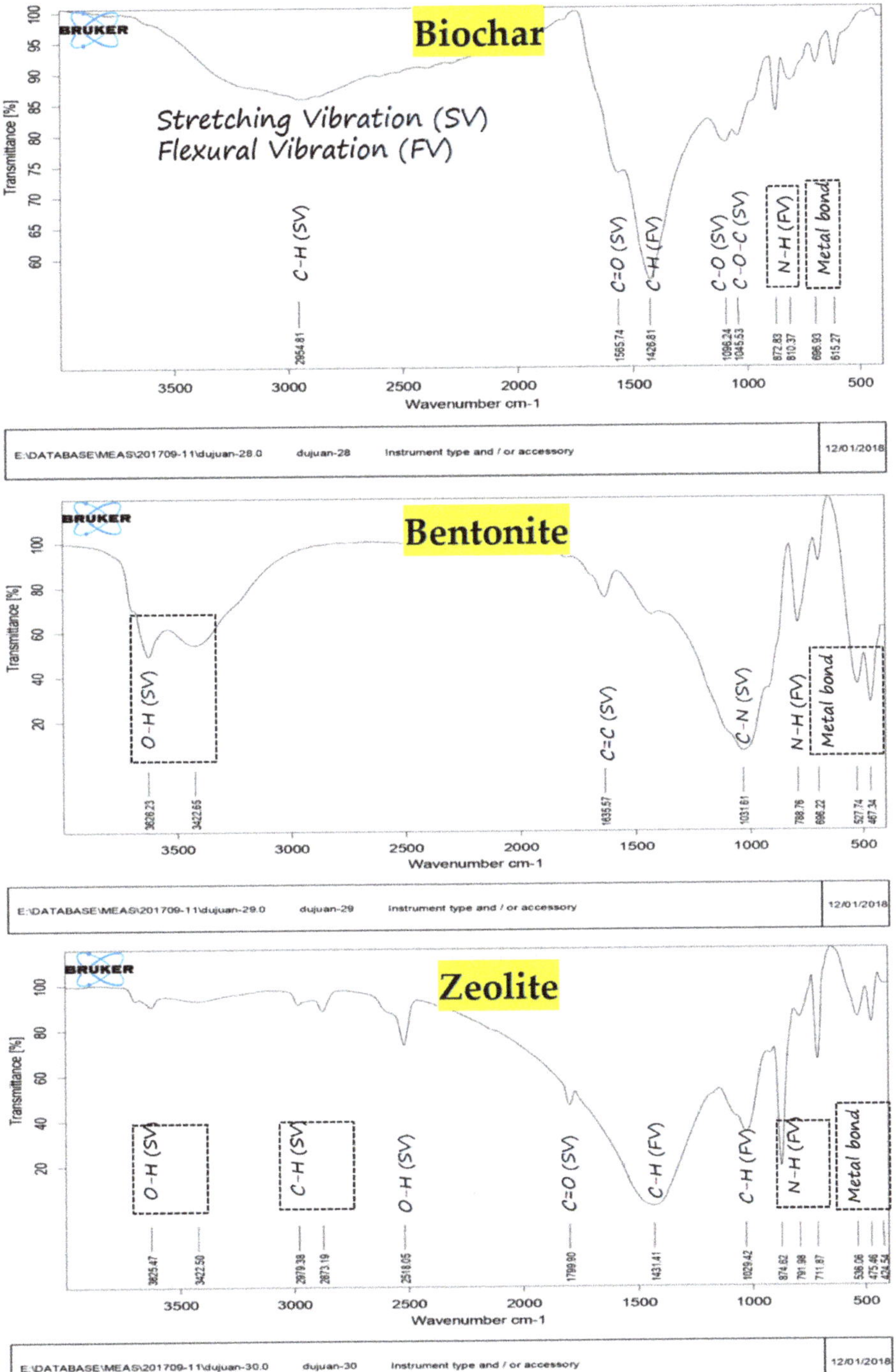


**Figure 3.** Fourier-transform infrared (FTIR) of biochar, bentonite, and zeolite.

### 3.1. Impact of Immobilizing Materials on Fresh and Dry Biomass and Chlorophyll Content by Pakchoi

The fresh and dry biomasses of the pakchoi crop were increased with the application of biochar and bentonite in smelter- and mine-polluted soils. The maximum fresh and dry biomasses (total root and shoot) of pakchoi were observed at 28.21 and 18.43% for smelter soil and 61.96 and 67.90% for mine soil with the application of zeolite compared with other treatments (Figure 4a,b). It has been observed that biochar, bentonite, and zeolite have positive impacts on pakchoi fresh and dry biomasses, but zeolite shows the highest potential because of its enriched in mineral–organic mixture, carbon, ammonical nitrogen source, mesoporous structure, and large specific surface area, which can act to absorb organic/inorganic pollutants and enhance plant growth and yield. Soil pH increased with the addition of additives; however, it did not find any negative impacts on plant growth. Contin et al. [54] observed that ryegrass growth was improved with the addition of zeolite in the contaminated soil due to reduced Ni toxicity after treated soil. Meanwhile, soil pH was increased but did not affect plant growth. The application of zeolite as an amendment increased the chlorophyll SPAD values of aerial pakchoi leaves, ranging from 35.0 to 40.9 (1.17 folds) for smelter soil, and in the case of mine soil, chlorophyll SPAD values increased, ranging from 31.8 to 40.1 (1.26 folds), respectively (Figure 4c). Irfan et al. [55] stated that the chlorophyll content in the soot of maize increased with rising BC doses of 0.5, 1, 2, and 4% compared with the control treatment.

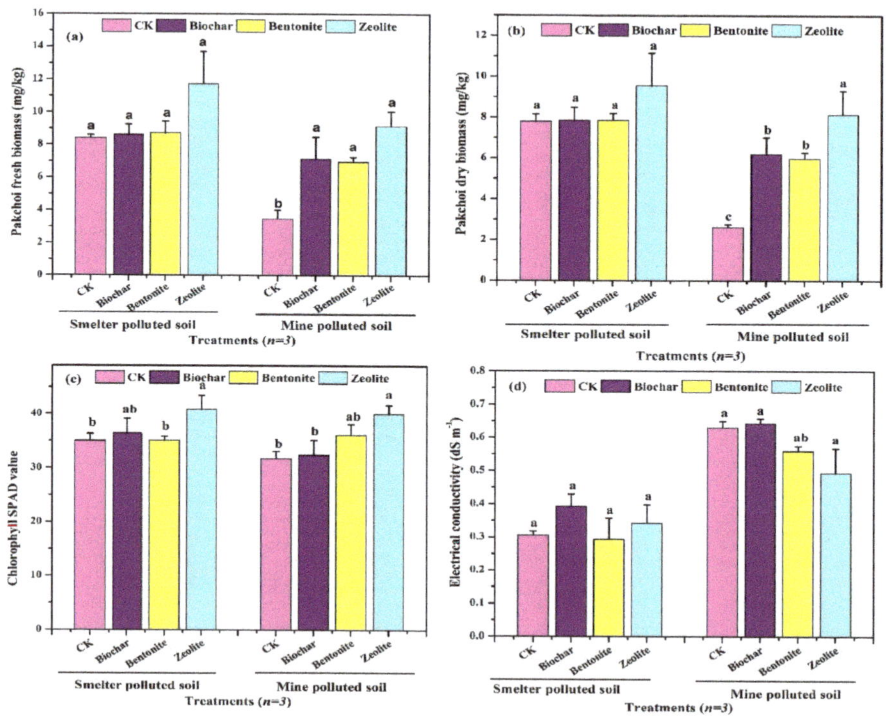


**Figure 4.** Impact of biochar, bentonite, and zeolite on pakchoi fresh biomass (**a**), pakchoi dry biomass (**b**), chlorophyll content (**c**), and electrical conductivity (**d**). The values in a given column followed by the same letter are not significantly different ($p < 0.05$) using highly significance difference (HSD) Tukey test.

*3.2. Impact of Immobilizing Materials on Soil Chemical Properties*

The addition of biochar, bentonite, and zeolite on the EC of smelter and mine soil was different. The maximum mean value and standard deviation of EC increased from 0.31 ± 0.1 to 0.39 ± 0.4 dS/cm in smelter soil with the application of biochar, but the application of bentonite decreased the EC, ranging from 0.31 ± 0.1 to 0.29 ± 0.06 dS/cm. The EC level in mine soil was enhanced from 0.63 ± 0.02 to 0.64 ± 0.01 with the application of zeolite, whereas the EC was reduced from 0.63 ± 0.02 to 0.49 ± 0.07 dS/cm over the control (Figure 4d). The application of soil additives increased pH mean values in smelter and mine soil; moreover, bentonite as an amendment potentially enhanced pH values to a great extent from 8.16 to 8.39 for smelter soil and 8.45 to 8.61 for mine soil compared to control soil (Figure 5a). Moeen et al. [56] found a rise in soil pH with the application of zeolite at a 5% dose.

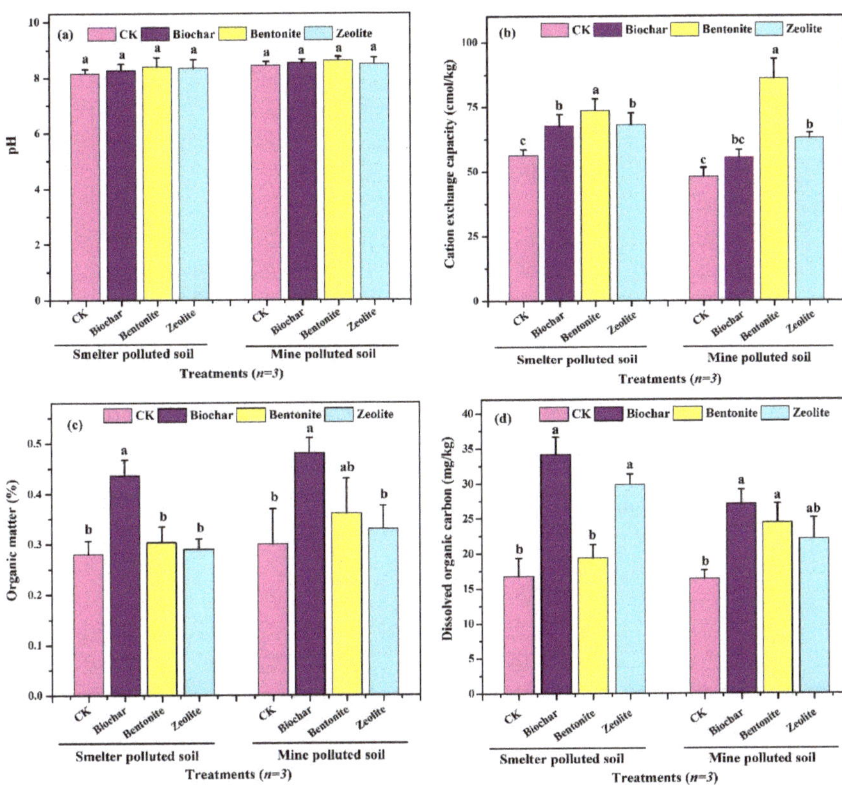


**Figure 5.** Impact of biochar, bentonite, and zeolite on pH (**a**), CEC (**b**), OM (**c**), and DOC (**d**). The values in a given column followed by the same letter are not significantly different ($p < 0.05$) using highly significance difference (HSD) Tukey test.

The highest mean values of CEC were observed, ranging from 56.4 to 73.6 (23.37%) in smelter soil and 48.0 to 85.7 (44.0%) in mine soil with the application of bentonite compared to treatments (Figure 5b). An increase in the mean value of OM and DOC was found to be highly significant, ranging from 0.28 to 0.44 (44.0%) and 0.30 to 0.48 (50.9%) for smelter soil and 16.8 to 34.2 (48.0%) and 16.4 to 27.1 (39.48%) for mine soil, respectively, with the application of biochar compared to other treatments (Figure 5c,d). The rise of soil pH has frequently been revealed to increase the proportion of OM, which may be responsible for increasing soil DOC [57–59]. Ali et al. [8] found an increase in soil pH with the application of rice straw-made biochar.

The maximum HA content in smelter- and mine-polluted soils was received by 57.14% and 30.3%, respectively, with a biochar application compared to other treatments (Figure 6a). The application of biochar as an amendment had shown a potential role to increase FA to a great extent of up to 61.54% in smelter soil, but the maximum FA in mine soil was up to 36.11% with the addition of zeolite compared to the control treatment (Figure 6b). Additionally, an increase in HA and FA in contaminated soil was observed with the addition of zeolite as an additive [60,61]. In addition, Shahbaz et al. [39] revealed that soil EC, pH, and OM were enhanced with the addition of rice straw-made biochar and zeolite in Ni-polluted soil. Irfan et al. [55] reported that the greatest increase in soil OM was recorded at 4% biochar compared to compost material. Moeen et al. [56] observed a rise in pH and cation exchange capacity in contaminated soil amended with zeolite addition. Feng et al. [40] exposed that the mixture of biochar and zeolite indicated the best effectiveness in variable soil pH and OM, thus enhancing the pakchoi biomass.

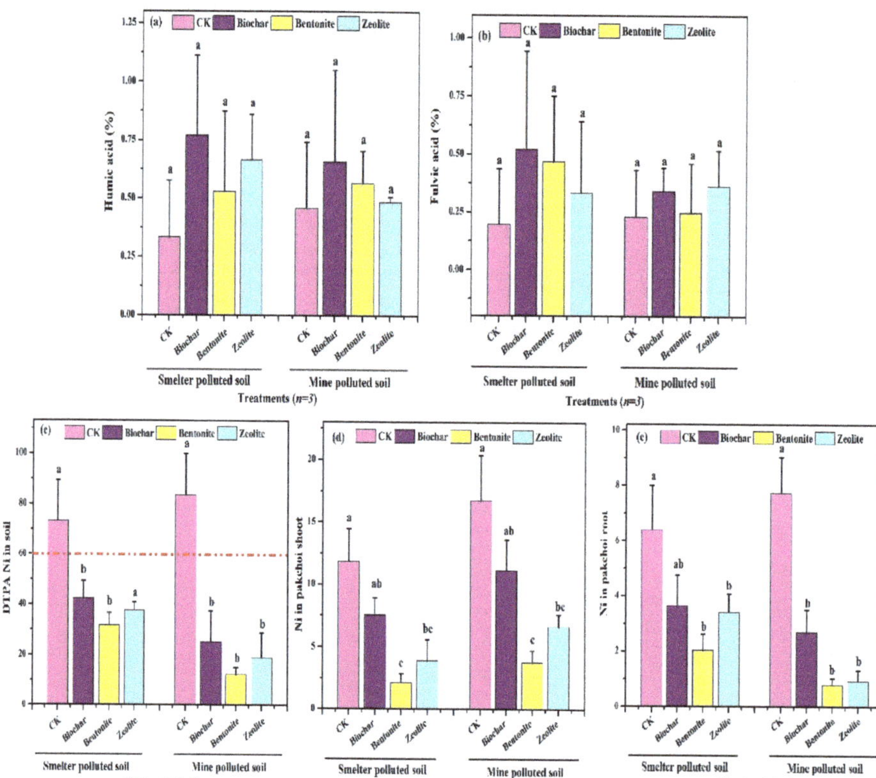


**Figure 6.** Impact of biochar, bentonite, and zeolite on humic acid (**a**), fulvic acid (**b**), DTPA-Ni (**c**), Ni in pakchoi shoot (**d**), and Ni in pakchoi root (**e**). The values in a given column followed by the same letter are not significantly different ($p < 0.05$) using highly significance difference (HSD) Tukey test.

### 3.3. Impact of Immobilizing Materials on Ni Immobilization

The addition of biochar, bentonite, and zeolite highly declined Ni mobility in smelter and mine soils. Moreover, the maximum Ni immobilization was observed at 76.8% for smelter soil and 85.38% for mine soil with the application of bentonite over the control (Figure 6c). It was observed that bentonite as an amendment was found to be highly effective for reducing Ni smelter- and mine-polluted soils compared to biochar and zeolite. This might be due to the rise of pH and CEC values; meanwhile, bentonite has a great surface area, maximum pore space, and high adsorption capacity. Shahbaz et al. [39] exposed that biochar 50% + zeolite 50% as a soil additive significantly fixed Ni in the soil. Ali et al. [8] found maximum Ni immobilization in polluted soil with the application of biochar and zeolite as amendments.

### 3.4. Impact of Immobilizing Materials on Ni Uptake in Shoots and Roots by Pakchoi

The uptake of Ni in the shoot and root by pakchoi plants was reduced with the application of ameliorants in smelter and mine soils. It was observed that the maximum reduction in Ni in the shoot and root was observed at 82.08% and 68.28% for smelter soil and 77.25% and 89.61% for mine-polluted soil with the application of bentonite compared to control soil (Figure 6d,e). Shahbaz et al. [39] reported that zeolite and biochar have the potential to reduce the accumulation of Ni in the root and shoot by wheat genotype.

### 3.5. Ni Fractions

The maximum reduction in the residual fraction of Ni in the smelter was observed at 64.17% with biochar and 63.86% with bentonite in mine-polluted soil. The highest decline of the oxidizable fraction of Ni was observed at 67.27% for smelter and 69.50% for mine soil with biochar addition. The addition of biochar and bentonite increased the reducible fraction of Ni up to 17.65% and 33.33%, but the highest reduction in the Ni reducible fraction was received at 85.71% with the addition of zeolite in smelter soil. The reducible speciation of Ni in mine soil was decreased by up to 48.19% with the application of bentonite compared to control soil. The highest reduction in the acid-soluble fractionation of Ni in smelter and mine soils was 70.15% and 74.64% with the incorporation of bentonite (Figure 7). The results of the present study indicated that the highest fixation of Ni was observed due to its conversion from available form to residual fractions. El-Naggar et al. [62] revealed that biochar was found to be a highly effective material for the stabilization of Ni in polluted soil. The reduction in acid-soluble speciation of Ni is because of the greater surface area and higher (ad)sorption capability of biochar and minerals [63]. Overall, the present study indicated that the residual, oxidizable, and acid-soluble fractions were reduced with the application of the studied amendments, whereas biochar and bentonite increased the reducible fraction of Ni in smelter soil. Also, Ali et al. [8] stated that biochar exposed a higher proportion of Ni in a reducible proportion by 26%–31.7%, respectively, compared with the control.

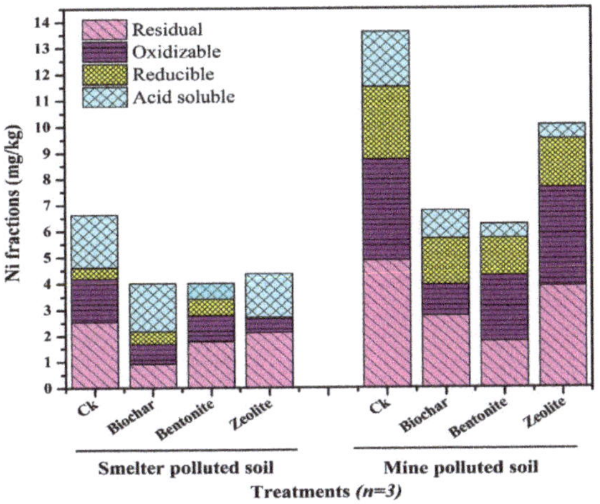


**Figure 7.** Impact of biochar, bentonite, and zeolite on Ni speciation in smelter- and mine-polluted soils.

### 3.6. Bioaccumulation and Translocation Factors

The bioaccumulation results indicated that the Ni bioaccumulation in the root by pakchoi from smelter soil was reduced, ranging from the control treatment 0.07–0.04 with biochar, 0.07–0.02 with bentonite, and 0.07–0.04 mg/kg with zeolite application (Table 2). The bioaccumulation of Ni in the root by pakchoi was reduced, ranging from the control treatment 0.8–0.02 with biochar, 0.8–0.008 with bentonite, and 0.08–0.01 mg/kg with zeolite application in mine-polluted soil (Table 2). It has been noted that bentonite as a clay mineral showed a high potential and was dominant in reducing the bioaccumulation of Ni in smelter-polluted soil, and zeolite was found to be highly effective for reducing Ni accumulation by pakchoi plants in the mine-polluted soil compared with bioaccumulation factor <1. The results of the translocation factor indicated that the Ni concentration increased from the control treatment, ranging from 0.84 to 2.06 with biochar, but the Ni translocation in pakchoi was reduced from the control treatment 1.84–1.04 with bentonite and 1.84–1.05 with zeolite in smelter-polluted soil (Table 2). The translocation of Ni in the pakchoi plant increased, ranging from the control treatment 2.16–4.15 with biochar 2.16–4.75 with bentonite and 2.16–7.33 mg/kg in mine-polluted soil. The translocation factor results indicated that the Ni accumulation from pakchoi root to the aerial surface was observed compared with standard values > 1 with the application of biochar, bentonite, and zeolite in smelter- and mine-polluted soil. This could be due to the pakchoi plant being considered a leafy vegetable plant and a hyperaccumulator plant that has the potential to phytomanage the Ni in polluted soil (Table 2); our results are in line with [41,51,52].

**Table 2.** Nickel bioaccumulation and translocation indices in soil and pakchoi.

| Treatments | Bioaccumulation Factor (BAF) | | Translocation Factor (TF) | |
|---|---|---|---|---|
| | Smelter-Polluted Soil (mg/kg) | Mine-Polluted Soil (mg/kg) | Smelter-Polluted Soil (mg/kg) | Mine-Polluted Soil (mg/kg) |
| Ck | 0.07 | 0.08 | 1.84 | 2.16 |
| Biochar | 0.04 | 0.02 | 2.06 | 4.15 |
| Bentonite | 0.02 | 0.008 | 1.04 | 4.75 |
| Zeolite | 0.04 | 0.01 | 1.13 | 7.33 |

### 3.7. Heat Map Correlation of Studied Parameters

As shown in Figure 8, the heat map correlation was performed among pakchoi shoot and root fresh and dry biomasses, Ni in smelter and mine soils, Ni in pakchoi shoots and roots, chlorophyll SPAD values, EC, pH, OM, CEC, HA, and FA. The obtained results show that biochar, bentonite, and zeolite have a positive correlation on pakchoi shoot and root fresh and dry biomasses in smelter-polluted soil. In addition, zeolite as an amendment showed a highly positive correlation on chlorophyll SPAD values in smelter-polluted soil. Biochar showed a positive correlation with DOC content. Bentonite as an amendment showed positive applications, and biochar, bentonite, and zeolite showed a positive correlation with DTPA-Ni in smelter-polluted soil, which indicated Ni immobilization in smelter soil. The positive correlation Ni in the pakchoi shoot was observed with biochar in smelter-contaminated soil. The response of biochar, bentonite, and zeolite was positively correlated with fresh shoot and root and dry shoot and root biomass by pakchoi in mine-polluted soil. Furthermore, zeolite as an amendment was positively correlated with pakchoi chlorophyll SPAD values in mine-polluted soil. The application of biochar, bentonite, and zeolite in mine soil showed a positive correlation with pH and DOC; meanwhile, study amendments potentially immobilized Ni in soil and reduced their uptake by pakchoi. The addition of bentonite in mine soil indicated a highly significant correlation with CEC in mine-polluted soil, which was found to be a high candidate for the immobilization of Ni in mine-contaminated soil. Sharma and Raju [64] assessed the correlation between heavy metals and soil properties; as a result, they did not find any significant correlation between Ni with soil EC and pH from different sites. Bahadur et al. [65] noted that pH and OM were observed to be significantly associated with corn dry biomass; furthermore, the corn dry biomass and soil pH were positively correlated with EC.

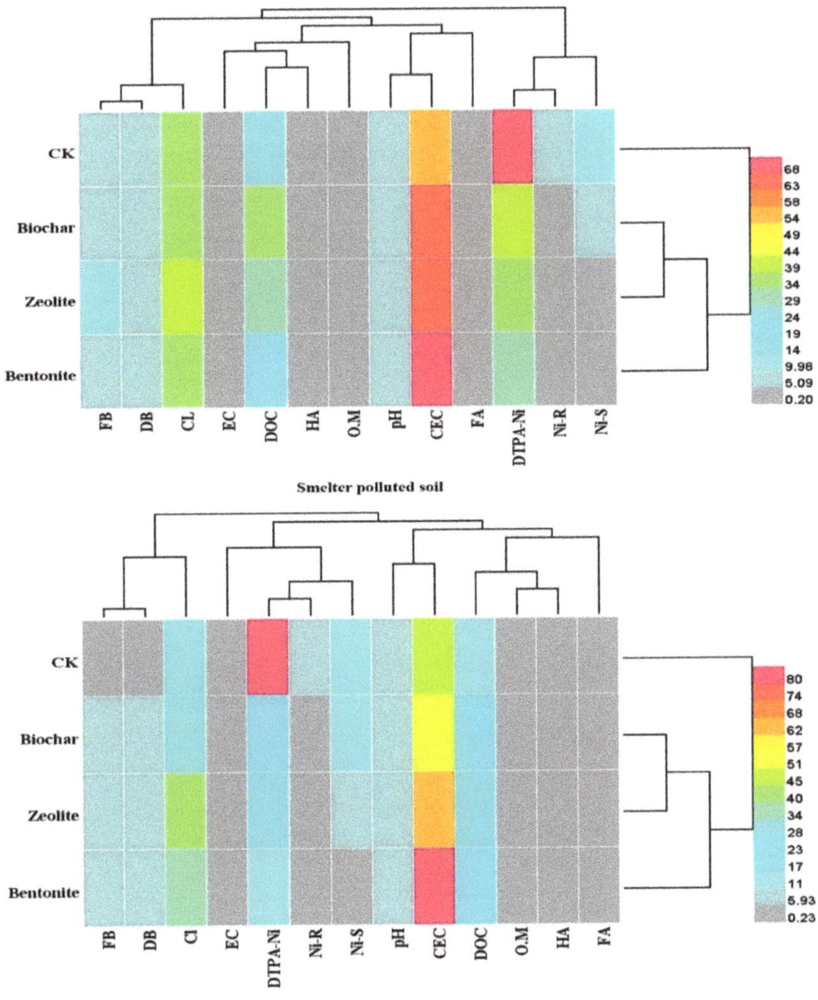


**Figure 8.** Impact of biochar, bentonite, and zeolite on heat map correlation between studied parameters in smelter- and mine-polluted soils. FB = fresh biomass, DB = dry biomass, Cl = chlorophyll, DOC = dissolved organic carbon, EC = electrical conductivity, HA = humic acid, OM = organic matter, pH = power of hydrogen, CEC = cation exchange capacity, FA = fulvic acid, DTPA-Ni = diethylene thiamine penta acetic acid–nickel, Ni-R = nickel in root, and Ni-S = nickel in shoot.

### 3.8. Principal Component Analysis

This statistical tool is used to analyze the correlation between the tested variables, in which the eigenvalues are presented in the given Figure 9. Here, 1–4 factors are more significant than others, as their values are more than one. This is clearly indicated in Figure 9. The loading plot (Figure 10) indicates that many variables are correlated to each other, including FA, DOC, Cl, and DB. They are similar components, and the rest of the others show different results. The PCA calculation shows that all data variables, DB, Cl, CEC, and DOC, correlate with PC1 and OM, and pH shows no PC1 correlation. PC2 accounts for 30% of the variance in the dataset. The treatments give positive relations in both PC2 and PC3, and the Ni root shows a negative correlation in PC2 and PC3. HA shows a correlation in PC4, and FA shows a negative correlation in PC4. DB is correlated with PC1 only, Cl with PC1 and 7 EC, pH and OM with PC2 only, and HA with PC 4 and 5. In all the studied parameters, PC 1 contributed 35%, PC2 contributed 24%, and PC3 contributed 12%, and approximately 60% was contributed by the first two PCs. The first three PCs contributed 71%, and almost 90% of the data are covered within PC5. Both loading and screen plots are presented, which show the relationships of the variables and the PC value contributions. The PC values are also presented in Table 3.

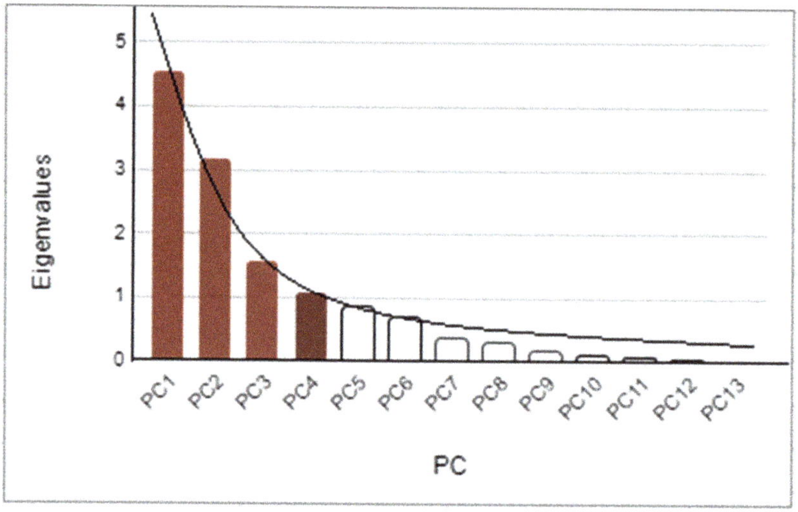


**Figure 9.** PC values in the screen plot.

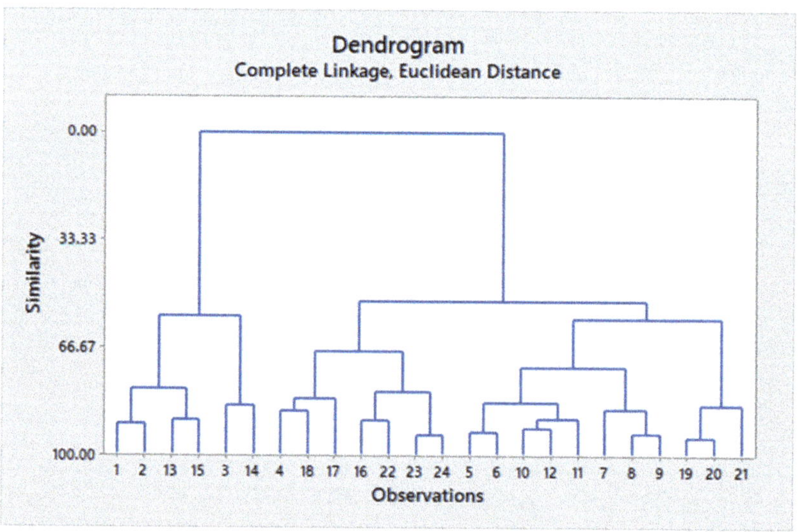


**Figure 10.** Cluster of different sites presented using Ni in comparison to different tested factors.

**Table 3.** Factorial analysis.

| Variables | Factor 1 | Factor 2 | Factor 3 |
|---|---|---|---|
| Trt | −0.40 | 0.759 | 0.395 |
| DB | 0.784 | −0.372 | −0.107 |
| CI | 0.678 | −0.275 | 0.228 |
| EC | −0.512 | 0.804 | 0.084 |
| pH | 0.003 | 0.760 | 0.168 |
| CEC | 0.755 | 0.082 | 0.269 |
| DOC | 0.561 | 0.291 | −0.627 |
| OM | 0.129 | 0.617 | −0.542 |
| HA | 0.273 | 0.357 | −0.377 |
| FA | 0.332 | 0.187 | −0.473 |
| DTPA-Ni | −0.719 | −0.556 | −0.236 |
| Ni Shoot | −0.903 | 0.047 | −0.253 |
| Ni Root | −0.788 | −0.471 | −0.270 |

A similar type of data fall in the same region and show similarity in behavior and form a cluster. When such observations of similar clusters combine, they form a dendrogram. The presented dendrogram is given in Figure 11. It shows two major groups that can also easily be observed from its score plot. These major groups comprise the smelter and mine areas. Further, the dendrogram is subdivided as per the factors analyzed. The score plot shows two groups of data; one belongs to smelter samples and the other to the mine, and it shows that smelter soil samples are similar to each other but different from mine samples and vice versa, which is why they are grouped separately. The smelter soil samples are towards the negative and zero values side, whereas the mine is towards the upper side of the system (Table 3 and Figure 12).

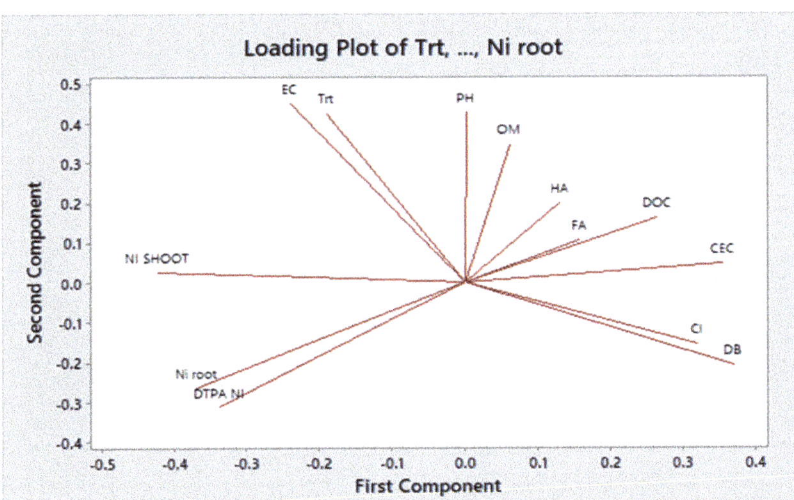


**Figure 11.** Loading plot of different variables using factorial analysis.

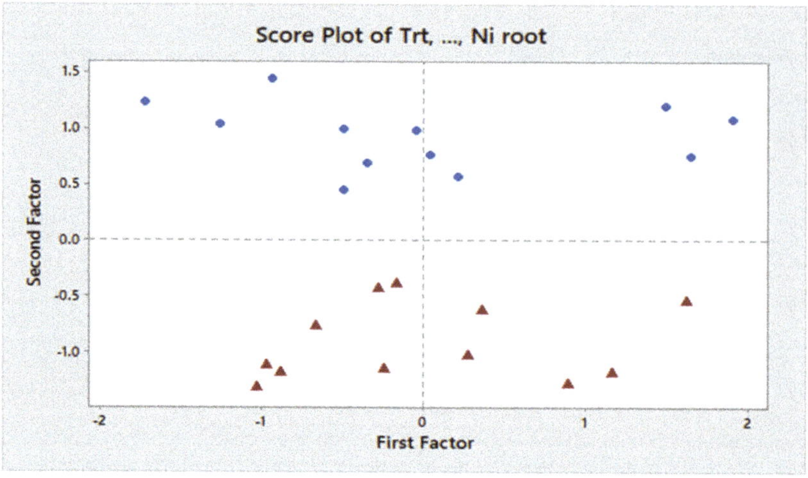


**Figure 12.** The score plot for Ni smelter data is in blue, and mine data are represented in red.

## 4. Conclusions

The results of the present trial revealed that the zeolite has greater potential to promote fresh and dry biomasses and chlorophyll SPAD values of pakchoi compared with biochar and bentonite in polluted soils. The application of bentonite was found to be highly effective for immobilizing Ni in smelter- and mine-contaminated soils, reducing their uptake by roots and shoots by pakchoi due to soil pH and CEC increases, respectively. Changes in the EC were observed in both soils with the application of soil material, but it did not possess any negative impacts on plant and Ni immobilization. The addition of bentonite in both soils increased the pH and CEC levels, which might be the main mechanism of Ni stabilization and reduce their absorption by pakchoi. The addition of biochar in smelter- and mine-contaminated soil enhanced the DOC, OM, and HA concentration compared with other treatments. The maximum FA was increased with the application of biochar in smelter soil and zeolite clay mineral as an amendment in mine soil. It has been suggested that zeolite and biochar have the potential to enhance plant growth, and bentonite clay mineral was found to be highly effective and dominant for reducing Ni in smelter (sandy loam) and zeolite clay minerals for the mine (loamy sand soil). The bioaccumulation of Ni in pakchoi root biomass was reduced with the application of bentonite in smelter-polluted and zeolite in mine-polluted soil compared with the control treatment. The translocation factor results were found to be higher and >1 with standard values after the application of additives. Long-term ex situ field experiments are suggested for the remediation of Ni in smelter- and mine-contaminated soils, decreasing their uptake by hybrid plants, soil biological properties, and health risk assessments of Ni in polluted soils with the integrated application of low-cost and environmentally friendly additives.

**Author Contributions:** Conceptualization, A.H.L.; methodology, A.H.L. and M.M.-H.; software, A.H.L. and A.U.; validation, V.V., S.V., and M.T.M.; formal analysis, A.A.; investigation, S.V.; resources, A.H.L. and M.M.-H.; data curation, A.H.L.; writing—original draft preparation, A.H.L.; writing—review and editing, S.R.A.; visualization, I.S. and A.A.T.; supervision, A.H.L.; project administration, A.H.L. and M.M.-H.; funding acquisition, M.M.-H. All authors have read and agreed to the published version of the manuscript.

**Funding:** This work was financially supported by the Shaanxi Province Qin Chuangyun "Scientists + Engineering" Team Building Project (2022KXJ-124). Furthermore, the research work was financed by the Ministry of Science and Higher Education of the Republic of Poland.

**Data Availability Statement:** Data will be made available upon request.

**Conflicts of Interest:** The authors declare that they have no known competing financial interests or personal relationships that could have appeared to influence the work reported in this paper.

## References

1. El-Naggar, A.; Ahmed, N.; Mosa, A.; Niazi, N.K.; Yousaf, B.; Sharma, A.; Sarkar, B.; Cai, Y.J.; Chang, C.X. Nickel in soil and water: Sources, biogeochemistry, and remediation using biochar. *J. Hazard. Mater.* **2021**, *419*, 126421. [CrossRef] [PubMed]
2. Lahori, A.H.; Ahmad, S.R.; Mierzwa Hersztek, M.; Afzal, M.; Afzal, A.; Bano, S.; Muhammad, M.T.; Aqsa, A.; Vambol, V.; Vambol, V. Comparative role of charcoal, biochar, hydrochar and modified biochar on bioavailability of toxic metals and enhance primary macronutrients in polluted alkaline soil. *Sci. Total Environ.* **2024**, *930*, 172810. [CrossRef] [PubMed]
3. Rouhani, A.; Skousen, J.; Tack, F.M. An Overview of Soil Pollution and Remediation Strategies in Coal Mining Regions. *Minerals* **2023**, *13*, 1064. [CrossRef]
4. Khan, Z.; Xianting, F.; Khan, M.N.; Khan, M.A.; Zhang, K.; Fu, Y.; Shen, H. The toxicity of heavy metals and plant signaling facilitated by biochar application: Implications for stress mitigation and crop production. *Chemosphere* **2022**, *308*, 136466. [CrossRef] [PubMed]
5. Sharma, P.; Kaur, J.; Katnoria, J.K. Assessment of spatial variations in pollution load of agricultural soil samples of Ludhiana district, Punjab. *Environ. Monit. Assess.* **2023**, *195*, 222. [CrossRef]
6. Sharma, A.; Nagpal, A.K. Contamination of vegetables with heavy metals across the globe: Hampering food security goal. *J. Food Sci.* **2020**, *57*, 391–403. [CrossRef]
7. Khan, W.D.; Ramzani, P.M.A.; Anjum, S.; Abbas, F.; Iqbal, M.; Yasar, A.; Ihsan, M.Z.; Anwar, M.N.; Baqar, M.; Tauqeer, H.M.; et al. Potential of miscanthus biochar to improve sandy soil health, in situ nickel immobilization in soil and nutritional quality of spinach. *Chemosphere* **2017**, *185*, 1144–1156. [CrossRef]
8. Ali, U.; Bashir, S.; Shaaban, M.; Zhou, X.; Gao, R.; Zhu, J.; Fu, Q.; Hu, H. Influence of various passivators for nickel immobilization in contaminated soil of China. *Environ. Eng. Sci.* **2019**, *36*, 1396–1403. [CrossRef]
9. Dotaniya, M.L.; Nagar, M.C. Can pressmud improve nickel availability in lead contaminated soils? *Natl. Acad. Sci. Lett.* **2023**, *46*, 47–50. [CrossRef]
10. Shukla, R.; Gopal, R. Excess Nickel Alters Growth, Metabolism, and Translocation of Certain Nutrients in Potato. *J. Plant Nutr.* **2009**, *32*, 1005–1014. [CrossRef]
11. Rattan, R.K.; Goswami, N.N. *Essential Nutrients and Their Uptake by Plants*; Fundamentals of Soil Science, Indian Society of Soil Science: New Delhi, India, 2002; pp. 309–332.
12. Mihandoost, M.; Rafati, M. Effect of biochar and vermicompost on the absorb of nickel metal from soil by cherry tomato (*Solanum lycopersicum* var. cerasiforme). *Water Soil Manage. Model.* **2023**, *3*, 157–170.
13. WHO. *Permissible Limits of Heavy Metals in Soil and Plants*; WHO: Geneva, Switzerland, 1996.
14. Hassan, M.U.; Chattha, M.U.; Khan, I.; Chattha, M.B.; Aamer, M.; Nawaz, M.; Ali, A.; Khan, M.A.; Khan, T.A. Nickel toxicity in plants: Reasons, toxic effects, tolerance mechanisms, and remediation possibilities a review. *Environ. Sci. Pollut. Res.* **2019**, *26*, 12673–12688. [CrossRef] [PubMed]
15. Beiyuan, J.; Li, J.S.; Tsang, D.C.W.; Wang, L.; Poon, C.S.; Li, X.D.; Fendorf, S. Fate of arsenic before and after chemical-enhanced washing of an arsenic-containing soil in Hong Kong. *Sci. Total Environ.* **2017**, *599–600*, 679–688. [CrossRef] [PubMed]
16. Chen, X.; Kumari, D.; Cao, C.J.; Plaza, G.; Achal, V. A review on remediation technologies for nickel-contaminated soil. *Hum. Ecol. Risk Assess. Int. J.* **2020**, *26*, 571–585. [CrossRef]
17. Saleem, I.; Lahori, A.H.; Mierzwa-Hersztek, M.; Afzal, A.; Muhammad, M.T.; Ahmed, M.S.; Vambol, V.; Vambol, S. The Application of MgO-Modified Biochars for the Immobilization of Ni, Cu, Pb, and Cr in Stone Crushing and Mining-Polluted Soil. *Agronomy* **2024**, *14*, 1423. [CrossRef]
18. Rajendran, S.; Priya, T.A.K.; Khoo, K.S.; Hoang, T.K.; Ng, H.S.; Munawaroh, H.S.H.; Karaman, C.; Orooji, Y.; Show, P.L. A critical review on various remediation approaches for heavy metal contaminants removal from contaminated soils. *Chemosphere* **2022**, *287*, 132369. [CrossRef]
19. Bokade, P.; Gaur, V.K.; Tripathi, V.; Bobate, S.; Manickam, N.; Bajaj, A. Bacterial remediation of pesticide polluted soils: Exploring the feasibility of site restoration. *J. Hazard. Mater.* **2023**, *441*, 129906. [CrossRef]
20. Derakhshan Nejad, Z.; Jung, M.C.; Kim, K.H. Remediation of soils contaminated with heavy metals with an emphasis on immobilization technology. *Environ. Geochem. Health* **2018**, *40*, 927–953. [CrossRef]
21. Xu, D.M.; Fu, R.B.; Wang, J.X.; Shi, Y.X.; Guo, X.P. Chemical stabilization remediation for heavy metals in contaminated soils on the latest decade: Available stabilizing materials and associated evaluation methods-A critical review. *J. Clean. Prod.* **2021**, *321*, 128730. [CrossRef]
22. Lam, N.T.; Song, S.X.; Dung, B.T.N.; Ahmed, S.R.; Lahori, A.H. Bioremediation of toxic metals and limnocharis flava plants growth in acidic polluted soil assisted by microbial inoculants. *Ecol. Quest.* **2024**, *35*, 1–20.
23. Koul, B.; Taak, P.; Koul, B.; Taak, P. Ex situ soil remediation strategies. In *Biotechnological Strategies for Effective Remediation of Polluted Soils*; Springer: Singapore, 2018; pp. 39–57. [CrossRef]
24. Srivastava, V.; Puri, M.; Srivastava, T.; Nidheesh, P.V.; Kumar, M.S. Integrated soil washing and bioreactor systems for the treatment of hexachlorocyclohexane contaminated soil: A review on enhanced degradation mechanisms, and factors affecting soil washing and bioreactor performances. *Environ. Res.* **2022**, *208*, 112752. [CrossRef] [PubMed]

25. Li, W.; Ni, P.; Yi, Y. Comparison of reactive magnesia, quick lime, and ordinary Portland cement for stabilization/solidification of heavy metal-contaminated soils. *Sci. Total Environ.* **2019**, *671*, 741–753. [CrossRef] [PubMed]
26. Srivastava, A.N.; Chakma, S. Assessment of in situ stabilization and heavy metal toxicity reduction of sugar mill pressmud through pilot scale composting. *Environ. Monit. Assess.* **2023**, *195*, 951. [CrossRef] [PubMed]
27. da Silva Medeiros, D.C.C.; Nzediegwu, C.; Benally, C.; Messele, S.A.; Kwak, J.H.; Naeth, M.A.; Ok, Y.S.; Chang, S.X.; El-Din, M.G. Pristine and engineered biochar for the removal of contaminants co-existing in several types of industrial wastewaters: A critical review. *Sci. Total Environ.* **2022**, *809*, 151120. [CrossRef] [PubMed]
28. Murtaza, G.; Ahmed, Z.; Eldin, S.M.; Ali, I.; Usman, M.; Iqbal, R.; Rizwan, M.; Abdel-Hameed, U.K.; Haider, A.A.; Tariq, A. Biochar as a green sorbent for remediation of polluted soils and associated toxicity risks: Critical Review. *Anal. Chem.* **2023**, *10*, 197. [CrossRef]
29. Usman, A.; Kuzyakov, Y.; Stahr, K. Effect of clay minerals on immobilization of heavy metals and microbial activity in a sewage sludge-contaminated soil (8 pp). *J. Soils Sediments.* **2005**, *5*, 245–252. [CrossRef]
30. Kumararaja, P.; Manjaiah, K.M.; Datta, S.C.; Shabeer, T.A. Potential of bentonite clay for heavy metal immobilization in soil. *Clay Res.* **2014**, *33*, 83–96.
31. Naveenkumar, T.; Backiyavathy, M.R.; Chitdeshwari, T.; Maheshwari, M.; Saraswathi, T.; Lakshmanan, A. Influence of zeolite on heavy metal immobilization in municipal solid waste compost contaminated soil. *J. Appl. Nat. Sci.* **2022**, *14*, 971–977.
32. Otunola, B.O.; Ololade, O.O. A review on the application of clay minerals as heavy metal adsorbents for remediation purposes. Environ. *Technol. Innov.* **2020**, *1*, 100692.
33. He, M.; Xu, Z.; Hou, D.; Gao, B.; Cao, X.; Ok, Y.S.; Rinklebe, J.; Bolan, N.S.; Tsang, D.C. Waste-derived biochar for water pollution control and sustainable development. *Nat. Rev. Earth Environ.* **2022**, *3*, 444–460. [CrossRef]
34. Cho, S.H.; Lee, S.; Kim, Y.; Song, H.; Lee, J.; Tsang, Y.F.; Chen, W.H.; Park, Y.K.; Lee, D.J.; Jung, S.; et al. Applications of agricultural residue biochars to removal of toxic gases emitted from chemical plants: A review. *Sci. Total Environ.* **2023**, *868*, 161655. [CrossRef] [PubMed]
35. He, L.; Zhong, H.; Liu, G.; Dai, Z.; Brookes, P.C.; Xu, J. Remediation of heavy metal contaminated soils by biochar: Mechanisms, potential risks and applications in China. *Environ. Pollut.* **2019**, *252*, 846–855. [CrossRef] [PubMed]
36. Sachdeva, S.; Kumar, R.; Sahoo, P.K.; Nadda, A.K. Recent advances in biochar amendments for immobilization of heavy metals in an agricultural ecosystem: A systematic review. *Environ. Pollut.* **2023**, *319*, 120937. [CrossRef] [PubMed]
37. Yang, X.; Tsibart, A.; Nam, H.; Hur, J.; El-Naggar, A.; Tack, F.M.G.; Wang, C.H.; Lee, Y.H.; Tsang, D.C.W.; Ok, Y.S. Effect of gasification biochar application on soil quality: Trace metal behavior, microbial community, and soil dissolved organic matter. *J. Hazard. Mater.* **2019**, *365*, 684–694. [CrossRef] [PubMed]
38. Mourgela, R.N.; Regkouzas, P.; Pellera, F.M.; Diamadopoulos, E. Ni(II) adsorption on biochars produced from different types of biomass. *Water Air Soil Pollut.* **2020**, *231*, 206. [CrossRef]
39. Shahbaz, A.K.; Ramzani, P.M.A.; Saeed, R.; Turan, V.; Iqbal, M.; Lewińskam, K.; Abbas, F.; Saqib, M.; Tauqeer, H.M.; Iqbal, M.; et al. Effects of biochar and zeolite soil amendments with foliar proline spray on nickel immobilization, nutritional quality and nickel concentrations in wheat. *Ecotox. Environ. Safe.* **2019**, *173*, 182–191. [CrossRef]
40. Feng, S.; Zhang, P.; Hu, Y.; Jin, F.; Liu, Y.; Cai, S.; Song, Z.; Zhang, X.; Nadezhda, T.; Guo, Z.; et al. Combined application of biochar and nano-zeolite enhanced cadmium immobilization and promote the growth of Pak Choi in cadmium contaminated soil. *Nano Impact* **2022**, *28*, 100421. [CrossRef]
41. Khan, Z.I.; Iqbal, S.; Ahmad, K.; Ashfaq, A.; Bashir, H.; Dogan, Y. Assessment of heavy metal content of wheat irrigated with wastewater in Sargodha, Pakistan: Implications for human health. *Trace Elem. Electroly* **2019**, *36*, 82–92. [CrossRef]
42. Turan, V. Calcite in combination with olive pulp biochar reduces Ni mobility in soil and its distribution in chili plant. *J. Int. Phytoremediat.* **2022**, *24*, 166–176. [CrossRef]
43. Gee, G.W.; Bauder, J.W. Particle-size analysis. In *Methods of Soil Analysis, Part 1. Physical and Mineralogical Methods, Agronomy Monograph No. 9*, 2nd ed.; Klute, A., Ed.; American Society of Agronomy/Soil Science Society of America: Madison, WI, USA, 1986; p. 383.
44. Antoniadis, V.; Alloway, B.J. The role of dissolved organic carbon in the mobility of Cd, Ni and Zn in sewage sludge-amended soils. *Environ. Pollut.* **2002**, *117*, 515–521. [CrossRef]
45. Lambrechts, T.; Couder, E.; Bernal, M.P.; Faz, A.; Iserentant, A.; Lutts, S. Assessment of heavy metal bioavailability in contaminated soils from a former mining area (la union, Spain) using a rhizospheric test. *Water Air Soil Pollut.* **2010**, *217*, 333–346. [CrossRef]
46. Donisa, C.; Mocanu, R.; Steinnes, E. Distribution of some major and minor elements between fulvic and humic acid fractions in natural soils. *Geoderma* **2003**, *111*, 75–84. [CrossRef]
47. Shahkolaie, S.S.; Baranimotlagh, M.; Dordipour, E.; Khormali, F. Effects of inorganic and organic amendments on physiological parameters and antioxidant enzymes activities in *Zea mays* L. from a cadmium-contaminated calcareous soil. *S. Afr. J. Bot.* **2020**, *128*, 132–140. [CrossRef]
48. Lu, R.K. *Analytical Methods of Soil Agrochemistry*; Chinese Agricultural Science and Technology Publishing House: Beijing, China, 1999.
49. Rauret, G.; Lopez-Sanchez, J.; Sahuquillo, A.; Rubio, R.; Davidson, C.; Ure, A.; Quevauviller, P. Improvement of the BCR three step sequential extraction procedure prior to the certification of new sediment and soil reference materials. *J. Environ. Monitor.* **1999**, *1*, 57–61. [CrossRef] [PubMed]

50. USEPA EPA. Method 3052: Microwave-assisted acid digestion of siliceous and organically based matrices. In *Test Methods for Evaluating Solid Waste*, 3rd ed.; United States Environmental Protection Agency: Washington, DC, USA, 1995.
51. Antonkiewicz, J.; Jasiewicz, C.; Koncewicz-Baran, M.; Sendor, R. Nickel bioaccumulation by the chosen plant species. *Acta Physiol. Plant.* **2016**, *38*, 40. [CrossRef]
52. Chen, H.; Yuan, X.; Li, T.; Hu, S.; Ji, J.; Wang, C. Characteristics of heavy metal transfer and their influencing factors in different soil–crop systems of the industrialization region, China. *Ecotox. Environ. Safe* **2016**, *126*, 193–201. [CrossRef]
53. *GB15618-1995;* State Environmental Protection Administration of China. Environment Quality Standard for Soils. Chinese National Standard Agency: Beijing, China, 1995.
54. Contin, M.; Miho, L.; Pellegrini, E.; Gjoka, F.; Shkurta, E. Effects of natural zeolites on ryegrass growth and bioavailability of Cd, Ni, Pb, and Zn in an Albanian contaminated soil. *J. Soils Sediments* **2019**, *19*, 4052–4062. [CrossRef]
55. Irfan, M.; Mudassir, M.; Khan, M.J.; Dawar, K.M.; Muhammad, D.; Mian, I.A.; Ali, W.; Fahad, S.; Saud, S.; Hayat, Z.; et al. Heavy metals immobilization and improvement in maize (*Zea mays* L.) growth amended with biochar and compost. *Sci. Rep.* **2021**, *11*, 18416. [CrossRef]
56. Moeen, M.; Qi, T.; Hussain, Z.; Ge, Q.; Maqbool, Z.; Jianjie, X.; Kaiqing, F. Use of zeolite to reduce the bioavailability of heavy metals in a contaminated soil. *J. Ecol. Eng.* **2020**, *21*, 186–196. [CrossRef]
57. Temminghoff, E. Chemical Speciation of Heavy Metals in Sandy Soils in Relation to Availability and Mobility. Ph.D. Thesis, Wageningen Agricultural University, Wageningen, The Netherlands, 1998.
58. Bolan, N.S.; Adriano, D.C.; Mani, P.A.; Duraisamy, A. Immobilization and phytoavailability of cadmium in variable charge soils, II. Eff. lime Addit. *Plant Soil.* **2003**, *251*, 187–198. [CrossRef]
59. Lahori, A.H.; Mierzwa-Hersztek, M.; Rashid, M.; Kalhoro, S.A.; Memon, M.; Naheed, Z.; Ahmed, M.; Zhang, Z.Q. Residual effects of tobacco biochar along with different fixing agents on stabilization of trace elements in multi-metal contaminated soils. *J. Environ. Sci.* **2020**, *87*, 99–309. [CrossRef] [PubMed]
60. Burlakovs, J.; Klavins, M.; Osinska, L.; Purmalis, O. The impact of humic substances as remediation agents to the speciation forms of metals in soil. *APCBEE Procedia* **2013**, *5*, 192–196. [CrossRef]
61. Lahori, A.H.; Mierzwa-Hersztek, M.; Demiraj, E.; Sajjad, R.U.; Ali, I.; Shehnaz, H.; Aziz, A.; Zuberi, M.H.; Pirzada, A.M.; Hassan, K.; et al. Direct and residual impacts of zeolite on the remediation of harmful elements in multiple contaminated soils using cabbage in rotation with corn. *Chemosphere* **2020**, *250*, 126317. [CrossRef] [PubMed]
62. El-Naggar, A.; Shaheen, S.M.; Ok, Y.S.; Rinklebe, J. Biochar affects the dissolved and colloidal concentrations of Cd, Cu, Ni, and Zn and their phytoavailability and potential mobility in a mining soil under dynamic redoxconditions. *Sci. Total Environ.* **2018**, *624*, 1059. [CrossRef]
63. Shen, Z.; Zhang, Y.; McMillan, O.; Jin, F.; Al-Tabbaa, A. Characteristics and mechanisms of nickel adsorption on biochars produced from wheat straw pellets and rice husk. *Environ. Sci. Pollut. Res.* **2017**, *24*, 12819. [CrossRef]
64. Sharma, M.R.; Raju, N.S. Correlation of heavy metal contamination with soil properties of industrial areas of Mysore, Karnataka, India by cluster analysis. *Int. J. Environ. Sci.* **2013**, *2*, 22–27.
65. Bahadur, D.S.; Ahmed, S.R.; Lahori, A.H.; Hussain, T.; Alvi, S.K.; Shafique, S.; Fatima, S.; Vambol, V.; Mierzwa-Hersztek, M.; Hinduja, P.; et al. Novel Fuller Earth, Rock Phosphate, and Biochar for Phytomanagement of Toxic Metals in Polluted Soils. *Agriculture* **2022**, *12*, 1216. [CrossRef]

minerals

MDPI

Article

# Five Years of Leaching Experiments to Evaluate Land Spreading of a Modified Bauxite Residue Before and After Treatment of Acid Mine Drainage: Sand or Soil Capping and Revegetation

**Patricia Merdy [1,*], Alexandre Parker [1], Chen Chen [1] and Pierre Hennebert [2]**

1 Université de Toulon, Aix Marseille Univ., CNRS, IM2NP, CEDEX 9, 83041 Toulon, France; salexandre.parker@orange.fr (A.P.); chen.83400@gmail.com (C.C.)
2 Ineris (French National Institute for Industrial Environment and Risks), BP 2, 60550 Verneuil-en-Halatte, France; pierre.hennebert@gmail.com
* Correspondence: patricia.merdy@univ-tln.fr

**Abstract:** The global generation of bauxite residue necessitates environmentally responsible disposal strategies. This study investigated the long-term (5-year) behavior of bauxite residue whose pH was lowered to 8.5, called modified bauxite residue (MBR), using lysimeters to test various configurations: raw MBR or used MBR (UMBR) previously applied for acid mine drainage remediation, sand or soil capping, and revegetation. Throughout the experiment and across all configurations, the pH of the leachates stabilized between 7 and 8 and their salinity decreased. Their low sodium absorption ratio (SAR) indicated minimal risk of material clogging and suitability for salt-tolerant plant growth. Leaching of potentially toxic elements, except vanadium, decreased rapidly after the first year to low levels. Leachate concentrations consistently remained below LD50 for *Hyalella azteca* and were at least an order of magnitude lower by the experiment's end, except for first-year chromium. Sand capping performed poorly, while revegetation and soil capping slightly increased leaching, though these were negligible given the low final leaching levels. Revegetated MBR shows promise as a suitable and sustainable solution for managing bauxite residues, provided the pH is maintained above 6.5. This study highlights the importance of long-term assessments and appropriate management strategies for bauxite residue disposal.

**Keywords:** red mud; toxic elements; environmental protection; remediation; lysimeters; acid mine drainage

## 1. Introduction

Bauxite residue, also known as red mud, is a byproduct of processing bauxite ore to produce alumina ($Al_2O_3$) through the Bayer process. This process involves heating bauxite in caustic soda (NaOH) at a high temperature and pressure to form sodium aluminate. The residual material consists of about 4% liquor and 55% solid, ranging from 0.3 to 2.5 tons per ton of $Al_2O_3$ produced. Worldwide, approximately 120 million tons of bauxite residue are generated annually. Primary concerns about bauxite residue disposal include salinity, alkalinity, physical properties, and the presence of potentially toxic metals, posing significant environmental hazards [1–3]. Therefore, it is crucial to explore methods for repurposing this waste and rehabilitating existing disposal sites.

The pH of the bauxite residue is alkaline, between 10 and 13, a consequence of the caustic soda used in the Bayer process. The elevated salinity and sodicity induce toxic effects on plants and animals. Consequently, if revegetation is contemplated, the neutralization of bauxite residue becomes a priority. Rai et al. [1] have explored various techniques, including the use of mineral acids, acidic waste, superphosphate and gypsum, coal dust, $CO_2$, silicate material, and seawater. Concerning their physical properties, bauxite residues exhibit low structural stability, rendering spreading deposits impermeable to rainwater percolation and hindering the development of vegetation root systems. This is attributed to their silty nature and the salinity that impedes particle aggregation [2]. In terms of mineralogical and chemical composition, bauxite residue varies across refinery plants but typically contains elevated concentrations of Fe oxides (goethite, hematite, magnetite) in addition to undissolved alumina, and Ti oxide (anatase, rutile, perovskite). The major elements in bauxite residue are Fe (20%–45%), Ti (5%–30%), Al (10%–22%), Si (20%–45%), and to a lesser extent Mn, Ca, Na, Cr, V, La, Sc, and Y [3]. Other elements are present at trace levels, such as As, Be, Cd, Cu, Gl, Pb, Hg, Ni, Th, U, V, Zn, Ce, Nd, and Sm. Some of these elements, like Sc and REE rare earths, are recoverable [4]. Non-metallic and non-metaloïd elements such as P and S may also be present [5]. Regarding radioelements, bauxite residue exhibits radioactive activity around 0.03 to 0.06 Bq $g^{-1}$ due to $^{238}U$ and 0.03 to 0.76 Bq $g^{-1}$ due to $^{232}Th$, both originating from the ore [5]. This may limit the use of bauxite residue as building material [6] or can be a source of danger in red mud dams [7].

Due to the substantial volumes of bauxite residue and the adverse environmental impact associated with simple disposal methods like ocean dumping, landfilling, or settling ponds, numerous efforts have been undertaken to explore environmentally sustainable and economically viable approaches for bauxite residue management. Reuse applications, including building materials, recovery of valuable chemicals, cement production, road and levee construction, and environmental remediation [8–12] have been proposed, potentially consuming more than 500,000 t $y^{-1}$ [13]. While this volume is substantial, it falls short of addressing the entirety of residues generated annually, necessitating consideration of alternative disposal methods. Landspreading has historically been a cost effective and widely employed disposal technique. However, a crucial condition for its application is to facilitate the successful revegetation of the spreading area [14,15], thereby mitigating the risk of wind and water erosion on the surface. Various techniques have been implemented to enhance revegetation, primarily focusing on reducing pH and improving the physical properties of the residue, resulting in a modified bauxite residue (MBR). Common practices include washing with a filter press [16], gypsum amendment [17–19], or the addition of coarse materials such as sand [20]. Even in areas where MBR deposits have been successfully revegetated through spreading, concerns persist regarding major elements and particularly toxic trace elements such as Cd, necessitating ongoing monitoring of leaching by groundwater.

Bauxite residue exhibits the capability to immobilize elements that may be mobile in aqueous solutions, encompassing both cationic and anionic species. Examples include phosphate from used water [21] and potentially toxic metal cations in applications such as remediating acid mine tailings (AMT) or addressing acid mine drainage (AMD). Acid mine drainage forms when sulfides in AMT oxidize upon exposure to air and water, generating sulfated acidic solutions with a pH below 6, resulting in various environmental problems [22,23]. Conventionally, alkaline chemicals like CaO or calcite are introduced to neutralize AMD [24]. Interestingly, bauxite residue can serve the same purpose and can also be mixed with AMT to inert it and foster revegetation.

There have been many studies on the possibility of revegetation and ecological restoration of bauxite residue spreading. Some rely on the extrapolation of characterization or laboratory experiments such as column leaching [25]. Regarding studies in an outdoor environment, in situ or using field-scale lysimeters, most of these consider the properties of the spreading at the time of their installation or at a single moment after it [26], temporal monitoring over one year being already considered a long time [27]. However, it is imperative that the ecological relevance of spreading be evaluated over the long term. Here, we propose, for the first time, a study in an outdoor environment with temporal monitoring over 5 years.

In a prior study, we explored this process by examining the long-term behavior of a mixture comprising modified bauxite residue (MBR) and acid mine tailings (AMT) [28]. In the current study, our focus is directed towards delineating the impact of the design of MBR spreading on the leaching of potentially toxic elements, with or without revegetation, with a sand or a soil capping. We also characterized the leaching from a previously utilized MBR (UMBR), which had been employed in the depollution of acid mine drainage (AMD). Utilizing nine lysimeters established for a duration of five years, we systematically examined the leaching of sulfate and 12 potentially toxic elements in aqueous solutions from both MBR and UMBR.

## 2. Materials and Methods

### *2.1. Modified Bauxite Residue (MBR) and Used Modified Bauxite Residue (UMBR)*

All chemical reagents were provided by VWR-advantor company (Rosny-sous-Bois, France) at analytical purity. An experimental flow diagram is given in Figure 1. Bauxaline® from ALTEO (Gardanne, France) is bauxite residue washed and partially dried in a filter press. The MBR used in the present study is Bauxaline® treated with atmospheric $CO_2$, to which 5% gypsum was added in order to reduce the pH to 8.5 with precipitation of $CaCO_3$. Soluble $Na_2SO_4$ remained. The MBR has a variable composition depending on the ore. Table 1 gives the usual range of major and potentially toxic elements in the Bauxaline and the values for the MBR used in this study, which are for most metal and metalloids above the limits for admission to inert waste storage facilities (IWSF) or non-hazardous waste storage facilities (NHWSF) [29]. The UMBR used in this study is the MBR described above, which was used to bind contaminants from acid mine drainage rich in As, Pb, Cd, Cr and Zn [30].

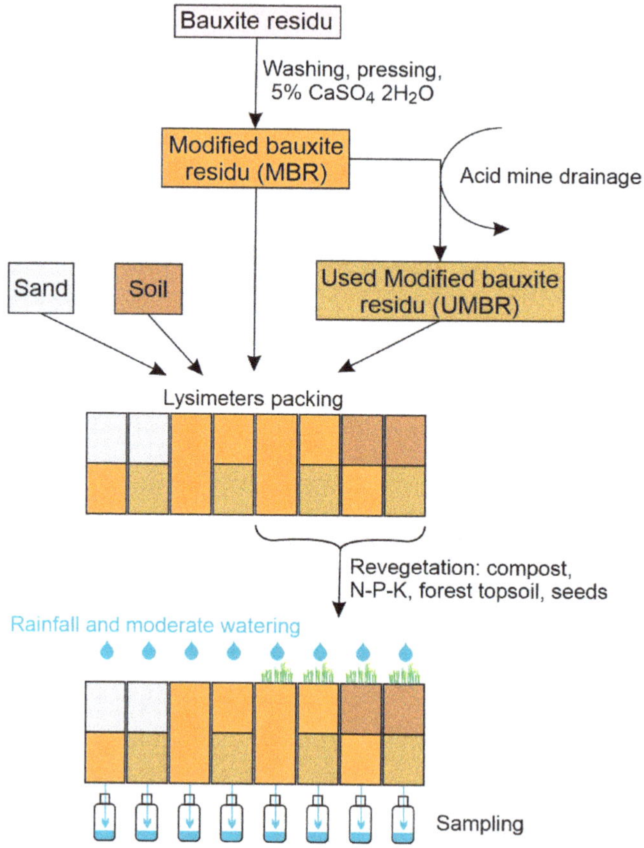


**Figure 1.** Experimental flow diagram.

**Table 1.** Usual range of major and potentially toxic elements in the Bauxaline® and composition of the Bauxaline® used in the present study. IWSF and NHWSF: concentration limits of leachable elements for admission to an inert waste storage facility and non-hazardous waste storage facility, respectively.

| | Major Elements (%) | | | Trace Elements (ppm) | | | |
|---|---|---|---|---|---|---|---|
| | Usual Range | Present Study | | Usual Range | Present Study | IWSF | NHWSF |
| Al | 1.6–8.1 | 7.9 | As | 10–200 | 18 | 0.5 | 2 |
| Ca | 1.4–4.8 | 4.3 | Cd | <0.5–10 | 0.8 | 0.04 | 1 |
| Fe | 21.0–38.0 | 32.2 | Co | 1–75 | 35 | | |
| Na | 1.5–3.0 | 3.0 | Cr | 200–2000 | 1638 | 0.5 | 10 |
| Si | 0.9–3.9 | 3.3 | Hg | <1.5–2 | 0.2 | | |
| Ti | 1.8–5.4 | 6.0 | Ni | 5–50 | 18 | 0.4 | 10 |
| | | | Pb | 10–100 | 42 | 0.5 | 10 |
| | | | Se | <1.5–50 | <6 | 0.1 | 0.5 |
| | | | V | 200–1500 | 968 | | |
| | | | Zn | <20–500 | 115 | 4 | 50 |

Two pilot trials were conducted to treat the highly contaminated acid mine drainage (AMD) from Saint-Félix (pH 2.2, As 57 mg $L^{-1}$, Cd 1.05 mg $L^{-1}$, Zn 117 mg $L^{-1}$) using 50 kg each of granulated MBR as metal binding. The trials involved treating 30 L of AMD per kg of MBR with a residence time of one hour. The decontamination efficiencies achieved were 99.89% and 99.63% for the two trials, respectively. The loading of contaminants in the MBR, calculated from the difference between influent and effluent concentrations, was 1710 mg $kg^{-1}$ for As, 38 mg $kg^{-1}$ for Cd, and 4100 mg $kg^{-1}$ for Zn.

### 2.2. Lysimeter Conception and Management

Eight 40-L lysimeters (hand-made from large plastic containers) were employed in this study, with varying content, including MBR, UMBR, silicate sand, and soil, with a total thickness of around 44 cm, as depicted in Figure 2. Lysimeters #1 to #4 remained unvegetated, while lysimeters #5 to #8 underwent a revegetation process. Additionally, a larger 700-L lysimeter (#16) filled with 66 cm (1087 kg) of MBR was subjected to the same revegetation process, aiming to more accurately simulate real MBR storage conditions (Figure 1). This last lysimeter was started a year after the others. To ensure technical and financial feasibility, we prioritized the collection of dense time series data over an extended period, without repetition. Repetition would have necessitated either sparse time series data or a shorter experimental duration. In this context, the consistency in the evolution of various variables over time ensures the validity of the results. Additionally, the comparison of the nine lysimeters further reinforces the robustness of the findings.

**Figure 2.** Lysimeters experimental setup. Left: 40-L lysimeters, view and diagram of filling and revegetation; center and right: 700-L lysimeter, shown from the side and from above, respectively.

The revegetalization was made using *Dactylis glomerata* (orchardgrass) and *Onobrychis sativa* (common sainfoin). The revegetation procedure encompassed the introduction of compost (1% $w/w$ in the top 20 cm layer), N-P-K fertilizer, forest topsoil (1‰ $w/w$ in the upper 22 cm layer) in lysimeters #5 and #6, which had no soil capping, and seeds. To address trace element requirements of carbonate-rich soils, an additional soluble manganese ($MnSO_4$) was added in the second and third years. Additionally, an annual application of N-P-K fertilizer, matching the initial dose, was conducted. Lysimeters #5, #6, and #16 underwent an initial watering prior to seeding to facilitate seedbed desalination.

Autumn, winter and spring rains were sufficient for plant growth. During the summer months, a moderate watering was conducted to sustain plants in a vegetative state. This approach facilitated their swift recovery once rainfall resumed, all while preventing runoff. It is noteworthy that the unvegetated lysimeters underwent the same watering regimen.

In years 4 and 5 of the study, a supplement of fresh forest plant litter was introduced onto the topsoil. Lysimeters #5 to #8 received 5 g each, while lysimeter #16 received 15 g. This deliberate addition aimed to foster the colonization of decomposing organisms, as organic matter had accumulated on the surface, forming a cohesive mat.

Notably, the revegetation efforts proved successful for each lysimeter where it was implemented.

### 2.3. Sample Collection and Analysis

After each rainfall event, drainage samples were systematically collected, acidified with ultrapure acid, and then stored at 2 °C. Each year, an annual composite sample was created, proportional to the annual drained volumes of the raw samples. For each sample, measurements of pH, electrical conductivity, and redox potential were recorded (VWR phenomenal MD8000L instrument, VWR-advantor company, Rosny-sous-Bois, France). However, redox potential measurements were not used because field tests have shown that the measured redox value was likely to be modified within a few minutes after contact with the atmosphere at the lysimeter outlet. Major and trace element analyses were performed using ICP-AES and ICP-MS at the Eurofins-certified laboratory.

Each sample volume was quantified by weighing, allowing for the calculation of the liquid/solid (L/S) ratio in liters per kilogram of residue. This ratio represents the volume of the leachate divided by the mass of the residue. Concentration is given in mg $L^{-1}$, while quantity data are given in mg $kg^{-1}$ by multiplying the concentration by the sample L/S ratio. Cumulative quantities correspond to the sum of the quantities of each sample over the duration of the experiment.

There are no international standards for element concentrations in landfill leachates. Existing regulations vary by country and type of filing. As toxicity is a key determinant of the ecological safety of leachates, we used an exposure test using the freshwater amphipod *Hyalella azteca*, chosen for its adaptability to the salinity of some of the leachates [31]. The method used is that described by ISO 16303 [32].

To assess the risk of clay dispersion presented by leachate salinity, we calculated the SAR (sodium absorption rate) using the formula

$$\mathrm{SAR} = \frac{[\mathrm{Na}^+]}{\sqrt{\frac{[\mathrm{Ca}^{2+}]+[\mathrm{Mg}^{2+}]}{2}}}$$

where $[Na^+]$, $[Ca^{2+}]$ and $[Mg^{2+}]$ are expressed in mmolc $L^{-1}$ [33].

## 3. Results and Discussion

### 3.1. Effect of Sand or Soil Capping on MBR Leaching

This topic was addressed by the results of lysimeters #1 (22 cm of MBR covered by 22 cm of sand), #3 (unvegetated 44 cm of MBR), #5 (vegetated 44 cm of MBR), #7 (22 cm of MBR covered by 22 cm of vegetated soil) and #16 (vegetated 66 cm of MBR).

### 3.1.1. Hydraulic Properties

The cumulative volumes over the experimental time of water input and drainage, expressed in mm of water height (equivalent to L $m^{-2}$), the drainage/water input ratio and the L/S ratio (litre per kg of MBR) are given in Figure 3 and Table 2.

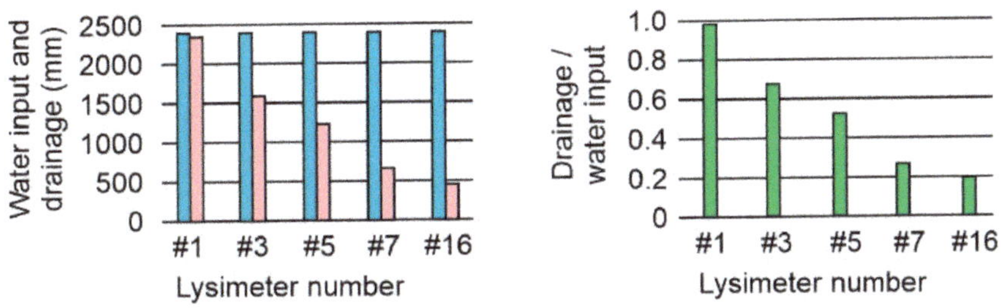


**Figure 3.** Cumulative water input (rainfall + watering) (blue columns), drainage (pink columns) and drainage/water input ratio (green columns) of lysimeters #1, #3, #5 and #16.

**Table 2.** Cumulated L/S ratio (L $kg^{-1}$ of MBR) of lysimeters #1, #3, #5, #7 and #16.

| | #1 | #3 | #5 | #7 | #16 |
|---|---|---|---|---|---|
| Year 1 | 2.68 | 1.16 | 0.88 | 1.02 | - |
| Year 1 to 2 | 3.85 | 1.45 | 1.02 | 1.08 | 0.09 |
| Year 1 to 3 | 5.12 | 1.85 | 1.33 | 1.60 | 0.26 |
| Year 1 to 4 | 6.67 | 2.22 | 1.46 | 1.72 | 0.30 |
| Year 1 to 5 | 8.89 | 2.99 | 2.31 | 2.48 | 0.46 |

A drainage-to-water input ratio of 99% was observed for lysimeter #1, suggesting that the sand capping significantly reduced evaporation. Conversely, lysimeters #3 and #5 exhibited ratios of 67% and 52% respectively. This reduction in drainage showed that revegetation increased evapotranspiration. The ratio was even lower for vegetated soil capping (26%), indicating that the soil capping improved water storage or evapotranspiration. The ratio was even lower, by only 20%, for lysimeter #16. Compared to lysimeter #5, the greater thickness of material in lysimeter #16 enabled the augmentation of water storage, thereby increasing the quantity of water available for evapotranspiration.

The potential for leaching of elements towards the depth is possibly higher when the L/S value is high, this will be commented on below (Section 3.1.3). Here, this value decreased with the presence of plant cover and with soil capping (Table 2). The values were nearly identical between lysimeters 5 (vegetated MBR) and 7 (vegetated soil capping), despite the quantity of MBR being half in the latter case, highlighting the effectiveness of soil capping.

### 3.1.2. pH and Salinity

During the first year of the experiment, the pH levels of leachate remained consistently above 8. Over the subsequent years, there was a gradual decrease, with pH values nearing 7 for lysimeters #1 and #3, and approximately 8 for lysimeters #5 and #16 in the last year (Figure 4). These pH variations align with the changes observed in electrical conductivity, which decreased from the beginning to the end of the experiment (see Figure 4). This indicates a decrease in the alkaline reserve in the soil solution. The latter was in equilibrium with the atmosphere, which explains the decrease in pH. At pHs of the order of 6.5 to 8, Al, Cr, V, Mo, and As exhibit minimal solubility in water, while Pb and Cu exhibit solubility below $10^{-5}$ mol $L^{-1}$ [34,35].

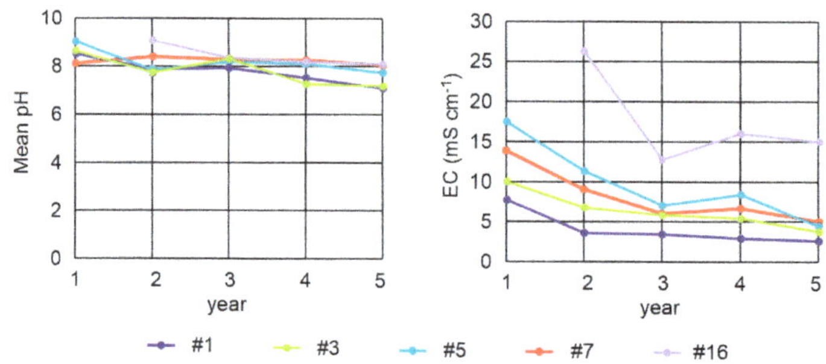


**Figure 4.** pH and electrical conductivity (EC) of lysimeters #1, #3, #5, #7 and #16.

The differences among lysimeters can be attributed to the cumulative L/S ratios, as shown in Table 2. Lower ratios indicate less solution passing through the lysimeter material, thus reducing the leaching of highly soluble salts that remained at higher concentrations at the end of the experiment. This resulted in higher conductivity and pH levels in the leachates at the end of the experiment. This was particularly clear for lysimeter #16.

3.1.3. Elements in Leachates

Figure 5 gives the temporal variation of the annual leaching per kg of MBR of the elements most likely to pose environmental hazards, such as Na and S, due to salinity or sulfation issues for plants and soils, as well as potentially toxic metals or metalloids. Data for all analyzed elements are provided in the Supplementary Materials Section. Figure 6 presents the cumulative leaching from each lysimeter for all analyzed elements over the entire experimental period (5 years for lysimeters #1 to #7 and 4 years for lysimeter #16).

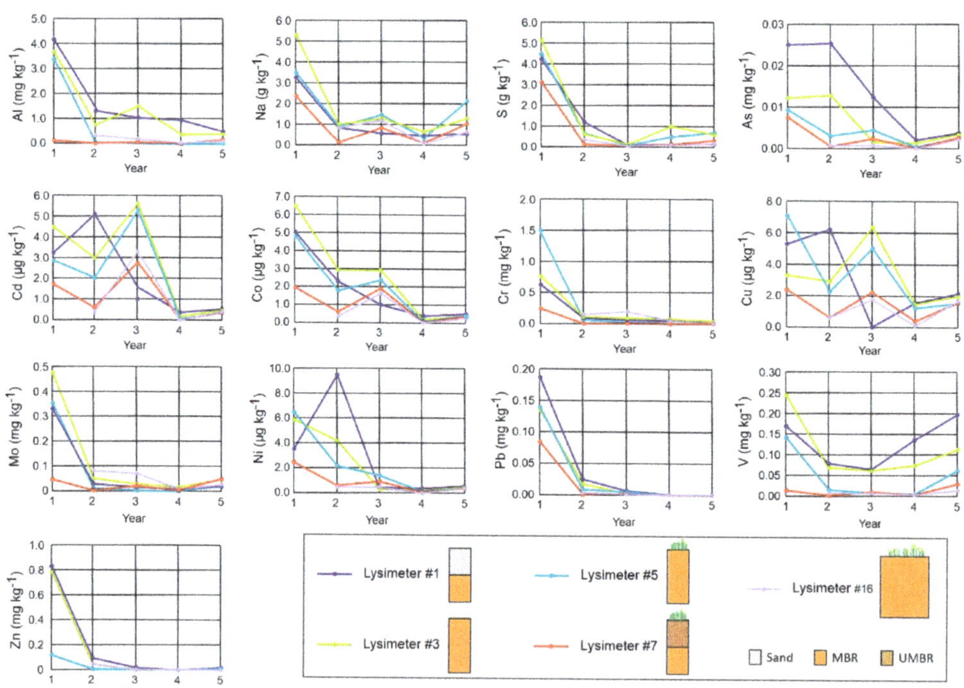


**Figure 5.** Variation of element leaching over time of some selected elements (Al, Na, S, As, Cd, Co, Cr, Cu, Mo, Ni, Pb, V, Zn) in mg $kg^{-1}$ of MBR. Sketch of lysimeters, see Figure 2.

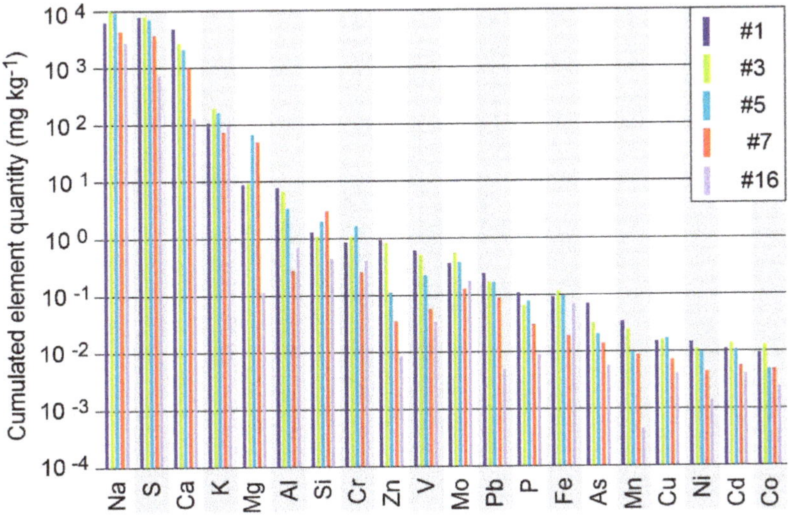


**Figure 6.** Cumulated element quantity (mg $kg^{-1}$ of MBR) leached by the lysimeters during the 5-year experiment for #1, #3, #5, and #7 and the 4-year experiment for #16.

For most elements in all lysimeters, leaching decreased over time, and for most elements, most of the decrease was observed in the first two years. This trend is primarily influenced by the quantity and availability of a soluble stock that is depleting. The slight increase in leaching between years 4 and 5 can likely be attributed to an intense rain event during year 5, where 72 mm of rainfall occurred following 54 mm of cumulative rainfall during the previous four days. This event saturated the lysimeters with water, facilitating diffusive transfer throughout their entire volume and likely temporarily lowering the material redox state, potentially leading to changes in species of the redox-sensitive elements such as As, Fe or V [36].

For most elements, leaching in lysimeters #1, #3 and #5 was higher than in lysimeters #7, and significantly higher than in lysimeter #16 (Figures 5 and 6). The latter difference can be attributed to the significantly lower L/S ratio in lysimeter #16 compared to other lysimeters. Lysimeter #7 had an L/S ratio similar to that of lysimeter #5 (Table 2); its lower element leaching was, therefore, due to soil capping. It should be noted that soil capping reduces the concentration peak observed in the first year of the experiment. Identifying the processes responsible for the better retention of leachable metals would require a dedicated study; however, one hypothesis is the production of non-leachable complexing organic matter in the soil, which could promote the immobilization of metals.

Average annual leachate concentrations are depicted in Figure 7. Consistent with annual leaching, concentrations for most elements were lower for lysimeters #7 and #16 compared to lysimeters #1, #3 and #5, respectively. Ca, Mg and P exhibited higher concentrations in all lysimeters during the final year compared to the first year of the experiment. The increase in Ca and Mg concentrations can be attributed to a pH decrease over the course of the experiment, as the solubility of these elements is highly pH sensitive. The increase in P concentrations is likely due to added fertilizer. Conversely, the concentrations of all other elements were lower in the final year compared to the first year, except for Si, V, Fe, As, Mn and Co in lysimeter #16, which exhibited higher concentrations in the final year. As previously mentioned, this can be due to an exceptional waterlogging period during the final year.

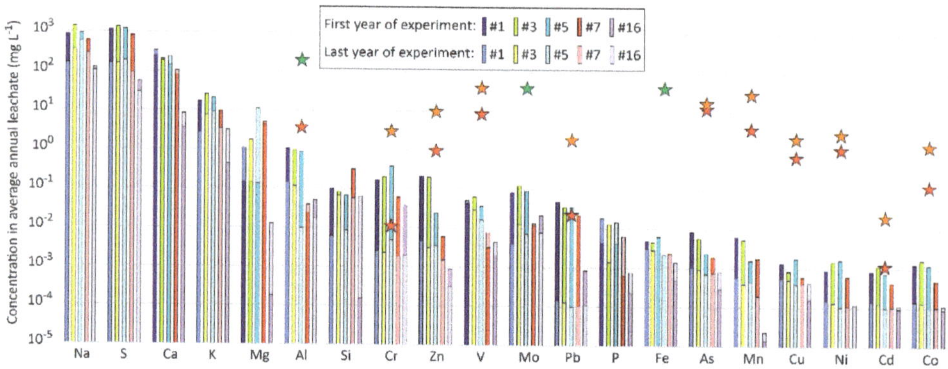


**Figure 7.** Concentrations in average annual leachate for the first year and the final year of the experiment. Stars indicate the LD50 concentrations for *Hyalella azteca* in soft freshwater (hardness 18, red stars) and hard freshwater (hardness 124, orange stars); green stars indicate that the LD50 concentration is over the star value [31].

Regarding the toxicity test carried out with *Hyalella azteca*, all observed concentrations remained below the LD50 level, except for Cr at the beginning of the experiment. Cd concentration was very close to the LD50 level at the beginning of the experiment. Concentrations of all elements remained under the LD50 values at the end of the experiment, especially knowing that the leachate water is hard (orange stars in Figure 7).

While the element concentrations in the leaching at the end of the experiment were low regardless of the experimental conditions, the quantities leached during the first years were lower for the vegetated soil capping configuration (#7) for all elements except Si.

Leachates contain high concentrations of Na that could cause the clay to disperse into the material underlying the spreading, leading to clogging of the porosity and making it impermeable [33,37]. The consequence would be the clogging of the spread materials and, therefore, a risk of death of vegetation, surface erosion by runoff and modification of the Eh conditions and, therefore, of the mobility of elements likely to change their oxidation state. SAR and EC values for the first and final years of the experiment are given in Figure 8. All points are plotted in the area, indicating no clay dispersion problems. All leachates, however, remained strongly saline with regard to plant growth.

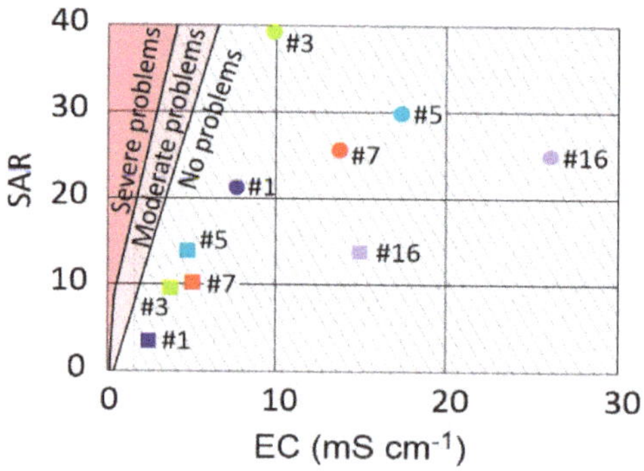


**Figure 8.** SAR and electrical conductivity (EC) of lysimeters #1, #3, #5, #7 and #16. Circles and squares give first and final-year values, respectively. Areas related to clay dispersion problems were drawn after Hanson et al. [38]. Hatched area represents high and very high salinity with regard to plant growth [39].

In summary, for all potentially toxic metals but V (Al, As, Cd, Co, Cr, Mo, Ni, Pb, Zn), element leaching was very low after 5 years regardless of the design, and concentration in leachates was an order of magnitude or lower than the LD50 toxicity level for *Hyalella azteca*. V leaching was relatively less reduced, but concentrations remained well below the toxicity threshold considered. This finding indicates efficient immobilization of these elements at final pH, whether by precipitation of oxides or adsorption on the active surfaces of clays or oxides.

### 3.1.4. Effect of Revegetation on MBR or UMBR Leaching

The impact of revegetation on element leaching was evaluated by comparing lysimeters #5 versus #3 (solely MBR) and #6 versus #4 (MBR over UMBR) (Figure 9). Overall, revegetation led to a decrease in leaching of most metals or metalloids investigated, except for Cr in both MBR-only and MBR/UMBR lysimeters, as well as to a small extent Cu in MBR-only lysimeter and Pb in MBR/UMBR lysimeter.

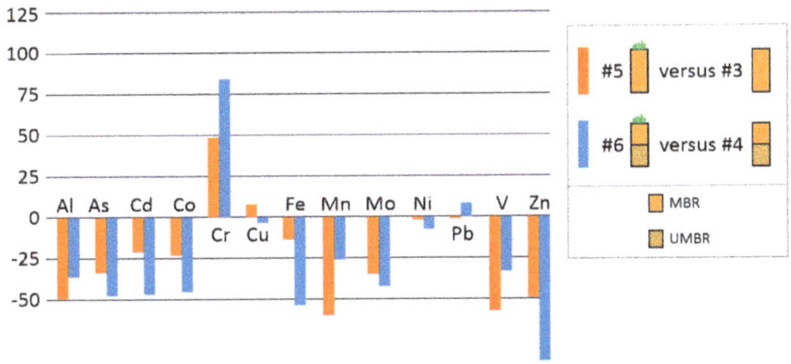


**Figure 9.** Element leaching from vegetated lysimeter (#5 and #6) as a % of the element leaching from the corresponding non-vegetated lysimeter (#3 and #4, respectively). Cumulative leaching during the 5 years of experimentation.

The mobility in leachates of the ionic species of a given metal $M$ with charge $c$ depends on the solubility of the metal-bearing minerals $\text{Min}_1$, $\text{Min}_2$, etc., its complexation with mobile dissolved organic matter (DOM), and its adsorption on solid surfaces, whether organic (solid organic matter, SOM) or mineral, $S_1$, $S_2$, etc. (e.g., oxides or clays). These interactions can be formally described by a set of equations such as the following, where the activities of the adsorbed species are calculated using the site density on the solid related to the solid/soil solution ratio and where $K_i$ is the stability constant of the corresponding reaction:

$$\begin{cases} \text{Min}_i + c\text{H}^+ + \cdots \Leftrightarrow M^c + \cdots & \text{with } K_{\text{Min}_i} = \dfrac{(M^c)(\ldots)^{\cdots}(\ldots)^{\cdots}\ etc.}{(M^c)\left(\text{H}^+\right)^c etc.} \\ M^c + \text{DOMH} \Leftrightarrow \text{DOM}M^{c-1} + \text{H}^+ & \text{with } K_{\text{DOM}M^c} = \dfrac{\left(\text{DOM}M^{c-1}\right)\left(\text{H}^+\right)}{(\text{DOMH})(M^c)} \\ M^c + \equiv \text{SOMH} \Leftrightarrow \equiv \text{SOM}M^{c-1} + \text{H}^+ & \text{with } K_{\equiv\text{SOM}M^c} = \dfrac{\left(\equiv\text{SOM}M^{c-1}\right)\left(\text{H}^+\right)}{(\equiv\text{SOMH})(M^c)} \\ M^c + \equiv \text{S}_i\text{H} \Leftrightarrow \equiv \text{S}_iM^{c-1} + \text{H}^+ & \text{with } K_{\equiv\text{S}_iM^c} = \dfrac{\left(\equiv\text{S}_iM^{c-1}\right)\left(\text{H}^+\right)}{(\equiv\text{S}_i\text{H})(M^c)} \\ & \text{etc.} \end{cases}$$

Parametrizing and solving such an equation system requires the use of a geochemical model such as PHREEQC [40,41] and would be the subject of a further study.

Biological activity associated with vegetation produces organic matter that can form stable organometallic complexes with metals or metalloids [42–46]. For instance, the complexation constants of a typical dissolved organic matter (DOM), as defined in the MinteqA2 database [47], vary for different metals: Al (5.2), Cd (3.3), CrIII (15.2), CuII (4.9), FeIII (7.7), Ni (3.3), Pb (5.2), and Zn (3.5). However, DOM enhances the mobility of an element only if its complexation constant is sufficiently high to facilitate the element desorption from solid matter, soil organic matter (SOM), or minerals. In this study, most elements exhibited lower leaching in the presence of vegetation, suggesting that their complexation constants with the available DOM were not high enough to promote mobility. In other words, the elements generally remained bound to the solid matter, SOM, or minerals despite the presence of DOM.

Chromium showed the opposite trend to most other elements, with higher leaching in vegetated lysimeters. This behavior can be attributed to different processes, such as oxidation of $Cr^{III}$ to highly mobile $Cr^{VI}$ species [48] or lower adsorption constants of $Cr^{III}$ on solid surfaces than on DOM. It is noted that the Cr-DOM complexation constant given in the MinteqA2 database is particularly high (15.2). Further complementary studies would be necessary to fully elucidate this behavior.

### 3.2. *Immobilization of Potentially Toxic Metals by UMBR That Has Depolluted Acid Mine Drainage*

This topic was addressed by the results of lysimeters #2 (22 cm of UMBR covered by 22 cm of sand), #4 (22 cm of UMBR covered by 22 cm of MBR), #6 (22 cm of UMBR covered by 22 cm of MBR), and #8 (22 cm of UMBR covered by 22 cm of vegetated soil) (Figure 2).

#### 3.2.1. Hydraulic Properties

The drainage/water input ratio (Figure 10) as well as the cumulated L/S ratio (Table 3) for lysimeters #2, #4, #6, and #8 were almost identical to those observed for lysimeters #1, #3, #5, and #7, respectively, leading to the same observation as for the latter lysimeters: sand capping limited evaporation, revegetation increased evaporation and reduced drainage, soil capping improved water storage or evapotranspiration.

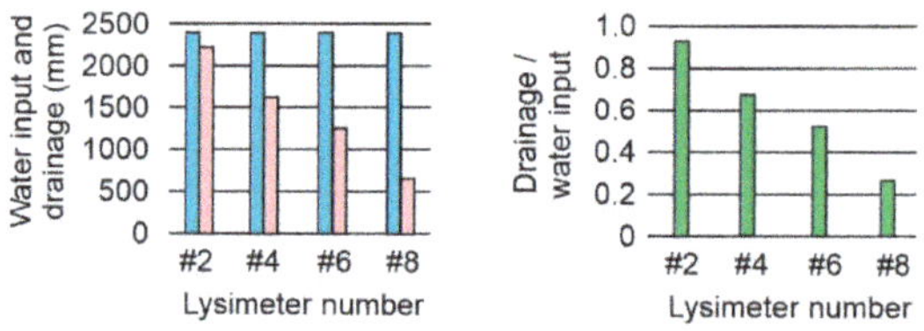


**Figure 10.** Water input (rainfall + watering) (blue columns), drainage (pink columns) and drainage/water input ratio (green columns) of lysimeters #2, #4, #6 and #8.

**Table 3.** Cumulated L/S ratio (L $kg^{-1}$ of MBR) of lysimeters #2, #4, #6 and #8.

| | #2 | #4 | #6 | #8 |
|---|---|---|---|---|
| Year 1 | 2.55 | 2.39 | 2.20 | 0.85 |
| Years 1 to 2 | 3.68 | 2.99 | 2.53 | 0.89 |
| Years 1 to 3 | 4.95 | 3.74 | 3.19 | 1.40 |
| Years 1 to 4 | 6.38 | 4.55 | 3.37 | 1.51 |
| Years 1 to 5 | 8.39 | 6.15 | 4.76 | 2.54 |

Lysimeters with the same coverage had very similar L/S ratios. Revegetation of the MBR cover (lysimeters #5 and #6) have reduced the drainage volume, without reaching the reduction observed for the revegetated soil capping (lysimeters #7 and #8).

#### 3.2.2. pH and Salinity

Compared to lysimeters #1, #3, #5 and #7, lysimeters #2, #4, #6 and #8 leachates showed slightly lower pHs and lower ECs at the beginning of the experiment, and almost identical pHs and ECs at the end of the experiment (Figures 4 and 11). This can be related to the fact that the more soluble salts had already been leached in the UMBR material during the acid mine treatment, resulting in pHs already below 8 and ECs already below 10 at the start of the experiment.

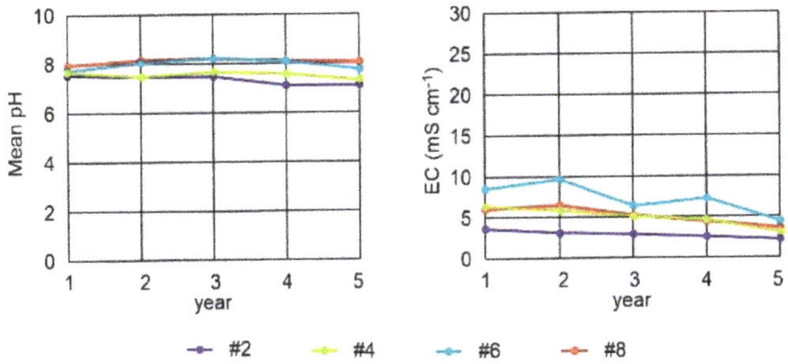


**Figure 11.** pH and electrical conductivity (EC) of lysimeters #2, #4, #6 and #8.

3.2.3. Effect of UMBR on Elements Leaching

Figure 12 shows the temporal variation of the annual leaching per kg of MBR or MBR+UMBR of potentially toxic metals or metalloids. Data for all analyzed elements are provided in the Supplementary Materials Section. In lysimeters with UMBR, the behavior of the elements is similar to that observed in lysimeters without UMBR, with leaching decreasing over time to reach very low values in the fifth year, with the exception of vanadium, as previously commented.

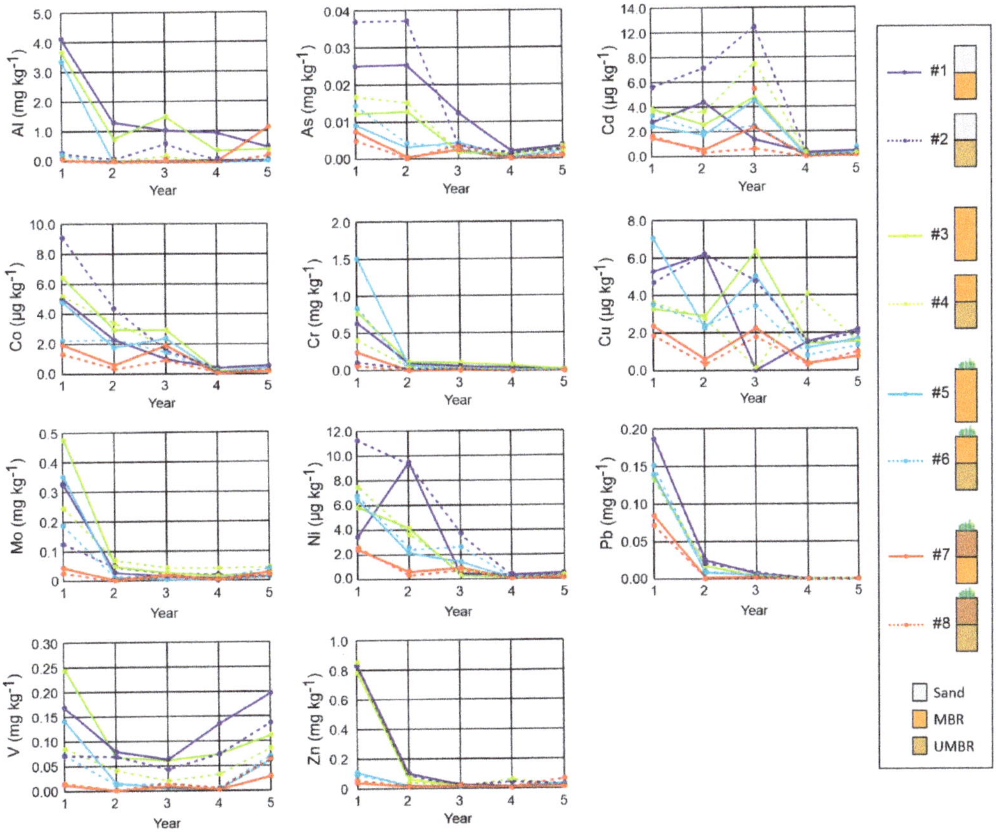


**Figure 12.** Variation of leaching over time of potentially toxic elements (Al, As, Cd, Co, Cr, Cu, Mo, Ni, Pb, V, Zn) in mg $kg^{-1}$ of (MBR+UMBR). Sketch of lysimeters, see Figure 2.

Figure 13 illustrates the variation in cumulative leaching between lysimeters with and without UMBR. There was no clear trend in the difference in leaching between lysimeters with and without UMBR. The presence of UMBR led to varying retention and leaching patterns for different elements: some elements, such as Al, Cr, and Mo, were more retained, while others, like As, Mn, and Ni, were more emitted. Additionally, the retention or leaching of certain elements varied depending on the lysimeter design: sand capping increased leaching for Cd, Co, Cu, Fe, and Ni, whereas soil capping led to higher leaching for Fe, Mn, V, and Zn. Interestingly, despite the higher As, Pb, Cr, and Zn contents in UMBR, lysimeters with UMBR did not consistently show higher leaching for these elements throughout the experiment. This indicates the effectiveness of UMBR in retaining these elements.

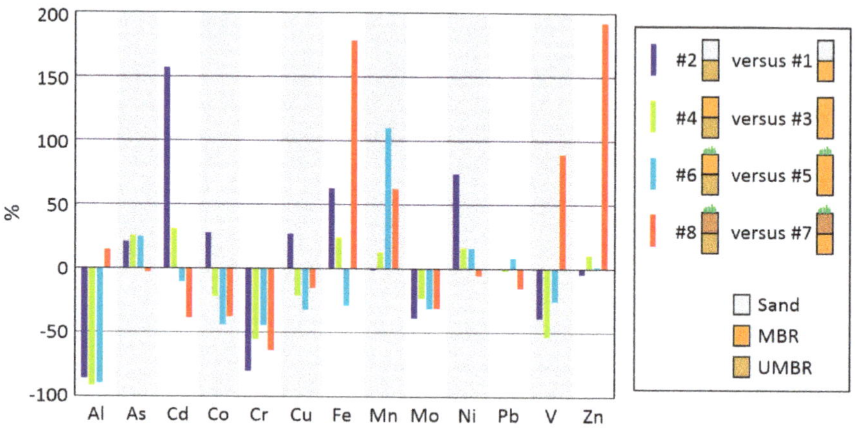


**Figure 13.** Element leaching from lysimeter with UMBR as a % of the element leaching from the corresponding lysimeter without UMBR. Cumulative leaching during the 5 years of experimentation.

As explained above, many parameters and mechanisms may be involved in the mobility of elements and their transfer into leachates. The treatment of AMD by UMBR led to a drop in pH; it could also have led to the precipitation of other mineral forms, modified the specific surfaces of the minerals or released sorption sites by the leaching of major mobile elements such as Ca or Mg. The precise determination of the mechanisms involved would require the fine characterization of the mineral phases present, the evaluation of their specific surfaces, in situ redox and pH monitoring in the lysimeters, the determination of the species present in the leachates and conceptual modeling.

Despite the observed variations, annual average concentrations remained below the LD50 values for all elements except Cr during the first year of the experiment, as previously mentioned.

## 4. Conclusions

Over the course of the 5-year experiment and across all configurations tested—raw, sand capping, soil capping, and revegetation—the pH of the leachates stabilized between 7 and 8, and their salinity gradually decreased. Although the salinity remained significant in the final year, ranging from 3 to 5 mS $cm^{-1}$, the SAR stayed well below the values that could cause clay dispersion and material clogging. Therefore, in all cases, the material remained suitable for the growth of plants compatible with the observed salinity.

Except for V, the leaching of potentially toxic elements from the modified bauxite residue (MBR), whether or not it was used to remediate acid mine drainage, rapidly decreased after the first year, reaching low levels. Except for Cr in the first year, the concentrations in the leachates always remained below the LD50 values and reached levels at least one order of magnitude lower than the LD50 by the end of the experiment. Among the different designs studied, soil capping and revegetation significantly reduced the leaching of elements. Soil capping, in particular, reduces the concentration peak observed in the first year of the experiment.

As a recommendation, in the climatic conditions of the test, which do not lead to soil acidification, spreading MBR, or an MBR previously used for AMD decontamination, with revegetalization after soil capping is the best environmental practice, which avoids the dispersion of dust and is not accompanied by the worry about leaching of potentially toxic elements. It will be necessary to check the composition of the residue before spreading, to verify that the contents of potentially toxic elements do not exceed the usual values as defined above, in particular vanadium. Under climatic conditions more conducive to soil acidification, for example, with high rainfall, the pH of the spreading should be controlled, not falling below 6.5, and adjusted if necessary by liming. Future research should establish content thresholds for the different potentially toxic elements in MBR-type matrices.

**Supplementary Materials:** The following supporting information can be downloaded at https://www.mdpi.com/article/10.3390/min14121210/s1, Table S1. Annual leachate emission (mg/kg of MBR).

**Author Contributions:** Conceptualization, P.M. and P.H.; methodology, P.H.; validation, P.M. and P.H.; investigation: P.M., P.H., C.C. and A.P.; writing—original draft preparation: P.M.; writing—review and editing, P.M. and P.H.; resources, P.H.; project administration, P.H.; funding acquisition, P.H. All authors have read and agreed to the published version of the manuscript.

**Funding:** Funding was provided by the BauGeste (also called Bauxaline Technologies) project (ADEME, Rhône-Mediterranean-Corsica Water Agency (AERMC agreement 2012/0547) and ALTEO) and the Bauxite+ project (ECCOREV-OHM).

**Data Availability Statement:** Data are given in Supplementary Materials.

**Conflicts of Interest:** The authors declare no conflicts of interest.

## References

1. Rai, S.; Wasewar, K.; Agnihotri, A. Treatment of alumina refinery waste (red mud) through neutralization techniques: A review. *Waste Manag. Res.* **2017**, *35*, 563–580. [CrossRef] [PubMed]
2. Newson, T.; Dyer, T.; Sharp, S. Effect of structure on the geotechnical properties of bauxite residue. *J. Geotech. Geoenviron. Eng.* **2006**, *132*, 143–151. [CrossRef]
3. Samal, S. Utilization of Red Mud as a Source for Metal Ions—A Review. *Materials* **2021**, *14*, 2211. [CrossRef] [PubMed]
4. Botelho Junior, A.B.; Espinosa, D.C.R.; Tenório, J.A.S. Characterization of bauxite residue from a press filter system: Comparative study and challenges for scandium extraction. *Mining Metal. Explor.* **2021**, *38*, 161–176. [CrossRef]
5. Gu, H.; Wang, N.; Liu, S. Radiological restrictions of using red mud as building material additive. *Waste Manag. Res.* **2012**, *30*, 961–965. [CrossRef]
6. Gu, H.; Wang, N.; Yang, Y.; Zhao, C.; Cui, S. Features of distribution of uranium and thorium in red mud. *Physicochem. Probl. Miner. Proc.* **2017**, *53*, 110–120. [CrossRef]
7. Kovács, T.; Sas, Z.; Jobbágy, V.; Csordás, A.; Szeiler, G.; Somlai, J. Radiological aspects of red mud disaster in Hungary. *Acta Geophys.* **2013**, *61*, 1026–1037. [CrossRef]
8. Liu, Y.; Naidu, R. Hidden values in bauxite residue (red mud): Recovery of metals. *Waste Manag.* **2014**, *34*, 2662–2673. [CrossRef]
9. Klauber, C.; Gräfe, M.; Power, G. Bauxite residue issues: II. options for residue utilization. *Hydrometallurgy* **2011**, *108*, 11–32. [CrossRef]
10. Verma, A.S.; Suri, N.M.; Kant, S. Applications of bauxite residue: A mini-review. *Waste Manag. Res.* **2017**, *35*, 999–1012. [CrossRef]
11. Kim, S.Y.; Jun, Y.; Jeon, D.; Oh, J.E. Synthesis of structural binder for red brick production based on red mud and fly ash activated using $Ca(OH)_2$ and $Na_2CO_3$. *Constr. Build. Mater.* **2017**, *147*, 101–116. [CrossRef]
12. Rivera, R.M.; Ulenaers, B.; Ounoughene, G.; Binnemans, K.; Van Gerven, T. Extraction of rare earths from bauxite residue (red mud) by dry digestion followed by water leaching. *Miner. Eng.* **2018**, *119*, 82–92. [CrossRef]
13. International Aluminium Institute. *Bauxite Residue Management: Best Practice*; IAI: London, UK, 2015; pp. 1–31.
14. Mishra, T.; Pandey, V.C.; Singh, P.; Singh, N.B.; Singh, N. Assessment of phytoremediation potential of native grass species growing on red mud deposits. *J. Geochem. Explor.* **2017**, *182*, 206–209. [CrossRef]
15. Mishra, T.; Singh, N.B.; Singh, N. Restoration of red mud deposits by naturally growing vegetation. *Int. J. Phytoremed.* **2017**, *19*, 439–445. [CrossRef]
16. Kinnarinen, T.; Lubieniecki, B.; Holliday, L.; Helsto, J.-J.; Häkkinen, A. Enabling safe dry cake disposal of bauxite residue by deliquoring and washing with a membrane filter press. *Waste Manag. Res.* **2015**, *33*, 258–266. [CrossRef]
17. Lindsay, W.L. *Chemical Equilibria in Soils*; John Wiley and Sons Ltd.: Hoboken, NJ, USA, 1979.

18. Barrow, N.J. Possibility of using caustic residue from bauxite for improving the chemical and physical properties of sandy soils. *Aust. J. Agric. Res.* **1982**, *33*, 275–285. [CrossRef]
19. Wong, J.W.C.; Ho, G.E. Use of waste gypsum in the revegetation on red mud deposits: A greenhouse study. *Waste Manag. Res.* **1993**, *11*, 249–256. [CrossRef]
20. Courtney, R.G.; Timpson, J.P. Reclamation of fine fraction bauxite processing residue. red mud) amended with coarse fraction residue and gypsum. *Water Air Soil Pollut.* **2005**, *164*, 91–102. [CrossRef]
21. Pradhan, J.; Das, J.; Das, S.; Thakur, R.S. Adsorption of Phosphate from Aqueous Solution Using Activated Red Mud. *J. Colloid Interface Sci.* **1998**, *204*, 169–172. [CrossRef]
22. Elghali, A.; Benzaazoua, M.; Taha, Y.; Amar, H.; Ait-khouia, Y.; Bouzahzah, H.; Hakkou, R. Prediction of acid mine drainage: Where we are. *Earth-Sci. Rev.* **2023**, *241*, 104421. [CrossRef]
23. Sarmiento, A.M.; Rodriguez-Perez, A.M.; Davila, J.M.; de la Torre, M.L. Effect of acid mine drainage on the mechanical properties of AISI 1020 carbon steel and AW6060 aluminium. *J. Mat. Res. Technol.* **2024**, *32*, 541–549. [CrossRef]
24. Kalin, M.; Fyson, A.; Wheeler, W.N. The chemistry of conventional and alternative treatment systems for the neutralization of acid mine drainage. *Sci. Total Environ.* **2006**, *366*, 395–408. [CrossRef]
25. Wu, Y.; Li, X.; Jiang, J.; William, H.; Zhu, F.; Xue, S. Integrating column leaching experiments and geochemical modelling to predict the long-term alkaline stability during erosion process for gypsum amended bauxite residue. *J. Environ. Manag.* **2021**, *289*, 112479. [CrossRef] [PubMed]
26. Jiang, Y.; Huang, S.; Zhu, F.; Guo, X.; Zhang, X.; Zhu, M.; Zhang, Y.; Xue, S. Long-term weathering difference in soil-like indicators of bauxite residue mediates the multifunctionality driven by microbial communities. *Sci. Total Environ.* **2023**, *890*, 164377. [CrossRef] [PubMed]
27. Scullett-Dean, G.; Stockwell, K.; Myers, L.; Nyeboer, H.; Santini, T.C. Accelerating soil formation in bauxite residue: A solution for long-term tailings management and storage. In Proceedings of the Mine Closure 2022: 15th International Conference on Mine Closure, Brisbane, Australia, 4–6 October 2022; Australian Centre for Geomechanics: Crawley, Australia, 2022; pp. 1027–1036. [CrossRef]
28. Merdy, P.; Parker, A.; Chen, C.; Hennebert, P. 5-year leaching experiments to evaluate a modified bauxite residue: Remediation of sulfidic mine tailings. *Environ. Sci. Pollut. Res.* **2023**, *30*, 96486–96498. [CrossRef] [PubMed]
29. European Communities. Council decision of 19 December 2002 establishing criteria and procedures for the acceptance of waste at landfills pursuant to Article 16 of and Annex II to Directive 1999/31/EC. *Off. J. Eur. Commun.* **2023**, *11*, 27–49. Available online: https://eur-lex.europa.eu/LexUriServ.do?uri=OJ:L:2003:011:0027:0049:EN:PDF (accessed on 12 October 2024).
30. Jacukowicz-Sobala, I.; Ociński, D.; Kociołek-Balawejder, E. Iron and aluminium oxides containing industrial wastes as adsorbents of heavy metals: Application possibilities and limitations. *Waste Manag. Res.* **2015**, *33*, 612–629. [CrossRef]
31. Borgmann, U.; Couillard, Y.; Doyle, P.; Dixon, D.G. Toxicity of sixty-three metals and metalloids to *Hyalella azteca* at two levels of water hardness. *Environ. Toxicol. Chem. Int. J.* **2005**, *24*, 641–652. [CrossRef]
32. *ISO 16303:2013*; Determination of Toxicity of Fresh Water Sediments Using *Hyalella azteca*. International Standardization Organization: Geneva, Switzerland, 2013. Available online: https://www.iso.org/obp/ui/#iso:std:iso:16303:ed-1:v1:en (accessed on 12 October 2024).
33. Rengasamy, P. Irrigation water quality and soil structural stability: A perspective with some new insights. *Agronomy* **2018**, *8*, 72. [CrossRef]
34. Cherfouh, R.; Lucas, Y.; Derridj, A.; Merdy, P. Metal speciation in sludges: A tool to evaluate risks of land application and to track heavy metal contamination in sewage network. *Environ. Sci. Pollut. Res.* **2022**, *29*, 70396–70407. [CrossRef]
35. Rocha, L.; Rodrigues, S.M.; Lopes, I.; Soares, A.M.V.M.; Duarte, A.C.; Pereira, E. The water-soluble fraction of potentially toxic elements in contaminated soils: Relationships between ecotoxicity, solubility and geochemical reactivity. *Chemosphere* **2011**, *84*, 1495–1505. [CrossRef] [PubMed]
36. Rinklebe, J.; Antoniadis, V.; Shaheen, S.M. Redox Chemistry of Vanadium in Soils and Sediments: Biogeochemical Factors Governing the Redox-Induced Mobilization of Vanadium in Soils. In *Vanadium in Soils and Plants*; CRC Press: Boca Raton, FL, USA, 2022; pp. 95–111. [CrossRef]
37. Cherfouh, R.; Lucas, Y.; Derridj, A.; Merdy, P. Long-term, low technicality sewage sludge amendment and irrigation with treated wastewater under Mediterranean climate: Impact on agronomical soil quality. *Environ. Sci. Pollut. Res.* **2018**, *25*, 35571–35581. [CrossRef] [PubMed]
38. Hanson, B.; Grattan, S.R.; Fulton, A. Agricultural salinity and drainage. In *Water Management Series Publication 3375*; University of California: Davis, CA, USA, 2006.
39. Wilcox, L.V. *Classification and Use of Irrigation Water (No. 969)*; U.S. Department of Agriculture: Washington, DC, USA, 1955.
40. Johnson, J.; Anderson, G.; Parkhurst, D. *Database Thermo.com.V8.R6.230, Rev 1.11*; Lawrence Livermore National Laboratory: Livermore, CA, USA, 2000; Volume 995.
41. Merdy, P.; Gamrani, M.; Montes, C.R.; Rezende Filho, A.T.; Barbiero, L.; Ishida, D.A.; Silva, A.R.C.; Melfi, A.J.; Lucas, Y. Processes and rates of formation defined by modelling in alkaline to acidic soil systems in Brazilian Pantanal wetland. *Catena* **2022**, *210*, 105876. [CrossRef]
42. Caporale, A.G.; Violante, A. Chemical processes affecting the mobility of heavy metals and metalloids in soil environments. *Curr. Pollut. Rep.* **2016**, *2*, 15–27. [CrossRef]

43. Merdy, P.; Guillon, E.; Aplincourt, M.; Duomonceau, J. Interaction of metallic cations with lignins. Part 1: Stability of iron (III), manganese (II) and copper (II) complexes with phenolic lignin model compounds: Coumaric, ferulic and sinapic acids and coniferyl alcohol. *J. Chem. Res.* **2000**, *2000*, 76–77. [CrossRef]
44. Merdy, P.; Lucas, Y.; Coulomb, B.; Melfi, A.J.; Montes, C.R. Soil organic carbon mobility in equatorial podzols: Soil column experiments. *Soil* **2021**, *7*, 585–594. [CrossRef]
45. Merdy, P.; Meunier, J.D.; Ziarelli, F.; Lucas, Y. Evidence of humic acid-aluminium-silicon complexes under controlled conditions. *Sci. Total Environ.* **2022**, *829*, 154601. [CrossRef]
46. Dobran, S.; Zagury, G.J. Arsenic speciation and mobilization in CCA-contaminated soils: Influence of organic matter content. *Sci. Total Environ.* **2006**, *364*, 239–250. [CrossRef]
47. Allison, J.D.; Brown, D.S.; Novo-Gradac, K.J. *MinteqA2/ProdefA2, a Geochemical Assessment Model for Environmental Systems, Database of Computer Programme*; Version 3.0; U.S. Environmental Protection Agency: Athens, GA, USA, 1991.
48. Pantsar-Kallio, M.; Reinikainen, S.P.; Oksanen, M. Interactions of soil components and their effects on speciation of chromium in soils. *Anal. Chim. Acta* **2001**, *439*, 9–17. [CrossRef]

minerals

MDPI

*Article*

# Assessment of Mercury Uptake by Plants in Former Cinnabar Mining Areas

**Milan Bauštein [1], Jiřina Száková [1,*], Luka Stefanović [1], Jana Najmanová [1], Jiřina Sysalová [2] and Pavel Tlustoš [1]**

[1] Department of Agroenvironmental Chemistry and Plant Nutrition, Czech University of Life Sciences in Prague, 16500 Prague, Czech Republic; milanbau@seznam.cz (M.B.); stefanovic@af.czu.cz (L.S.); najmanova@af.czu.cz (J.N.); tlustos@af.czu.cz (P.T.)

[2] AAS Laboratory, University of Chemistry and Technology, 16628 Prague, Czech Republic; jirina.sysalova@vscht.cz

* Correspondence: szakova@af.czu.cz

**Abstract:** Assessment of the plant's ability to take up mercury (Hg) from polluted soil was affected by location, plant family, and species in two former cinnabar mining areas in the Czech Republic. At each location, seven sampling points were marked out in the vicinity of former shafts and dumpsites connected to the mining activity, where representative soil samples and dicotyledonous plants were collected. The individual locations were characterized by specific plant communities, where, in most cases, different plant species were found within one family at both locations. The total Hg content in the soil, as well as gaseous elemental mercury ($GEM_{soil\text{-}air}$), confirmed elevated levels of this element in the mining-affected environment, with high variability of the data. The low Hg accumulation ability of plants, especially the low root–shoot translocation in most of the plant species, indicated the predominant occurrence of excluders. Among the families, the results showed the exceptional position of the Fabaceae family regarding soil Hg pollution, as the highest Hg content in both shoots and roots was determined for *Onobrychis viciifolia*. Therefore, the behavior of Fabaceae plants in polluted soil, the mechanisms of their tolerance to high Hg content, and their Hg accumulation ability deserve further research.

**Keywords:** mercury; soil; plant; accumulation; translocation

## 1. Introduction

Mercury (Hg) is among the most notorious and severe environmental pollutants. It is a toxic, persistent, and mobile contaminant [1]. Elevated Hg levels in the soil increase the risk to food security and the soil–plant–animal transport of this element [2]. Elevated Hg content in the soil can originate from natural (geogenic) as well as human-caused sources (anthropogenic). Geogenic Hg originates primarily from geological sources, such as volcanic eruptions, thermal springs, and surface metallic ores, whereas anthropogenic Hg sources primarily include coal-burning power plants and by-products of the mining process for metal ores (e.g., gold and cinnabar). In this context, artisanal and small-scale gold mining activities applying the amalgamation process are among the most serious sources of Hg pollution. In addition to direct Hg production, Hg can be released from other industrial activities, such as the manufacturing of cement, nonferrous metals, pig iron and steel, caustic soda, and waste disposal. Chemical industries, such as chlor-alkali plants, should also be taken into account [3]. However, mining and processing of Hg-bearing ores can result in hotspots of extremely Hg-polluted soils in mining areas.

An elevated soil Hg content poses a potential risk to other components of the ecosystem, especially through soil–plant transport and the accumulation of this element up the food chain. The toxicity, accumulation, and mobility of mercury depend on its form in the soil as well as the soil's physical, chemical, and biological properties. For example, organic forms of Hg, such as methylmercury (MeHg), are at least an order of magnitude more mobile than inorganic forms and are therefore more toxic and more readily bioaccumulated [4]. Fernandez-Martinez et al. [5] identified both humic and fulvic acids, as well as elemental Hg, as the primary variables controlling Hg methylation in soils. However, the Hg methylation rate in soils can also be affected by soil–plant relationships. Some bacteria can resist heavy metal contamination through chemical Hg transformation by reduction, oxidation, methylation, and demethylation [6]. The genetic system that evolved as the "mer operon" is the only well-known bacterial Hg resistance system allowing for the transformation of its toxic target into volatile non-toxic forms [7].

The soil physicochemical parameters and soil chemical composition contribute to Hg mobility and subsequent plant uptake of this element. Heeraman et al. [8] tested the potential effect of various soil amendments, such as lime, nitrogen (N), phosphorus (P), and organic matter, on plant growth and Hg uptake by *Vulpia myuros* growing in acidic mine soil. They confirmed that both liming (i.e., increasing soil pH to 6) and organic matter addition were effective measures to reduce Hg uptake by plants. In contrast, the addition of thiol ligands, such as ammonium thiosulphate ($(NH_4)_2S_2O_3$) and ammonium thiocyanate ($NH_4SCN$), enhanced Hg uptake by *Brassica juncea* plants grown in mine tailings originating from artisanal gold mines utilizing the amalgamation method for gold extraction [9]. Soil organic matter affects the speciation, solubility, mobility, and toxicity of Hg.

Many organic compounds have a high affinity for Hg due to their functional groups, such as hydroxyl-, carboxylic-, aromatic-, and S-containing ligands. Decreased soil Hg mobility, followed by suppressed plant uptake and the high sorption ability of soil organic matter for Hg, have been documented in organic-matter-amended soils [8,10]. Moreover, Hg mobility is affected by interactions between mineral and organic phases [11] as well as aggregate size [12]. Hg species, such as elemental mercury (Hg) and Hg–metal amalgams, are less toxic than soluble inorganic Hg species, such as mercury chloride ($HgCl_2$), due to their lower mobility in environmental processes [13]. Reis et al. [14] found a high percentage of immobile Hg in soils contaminated by mining activities due to a high sulfur content in these soils.

As reviewed by Beckers and Rinklebe [15], elevated Hg contents in plants are found in areas with high soil Hg contents, but no clear relationships have been observed between elevated Hg concentrations in plants and soils. The variability of the plant Hg content depends on the ability of the individual plant species to accumulate this element. An extremely wide range of plant Hg contents has been documented among plant species growing in extremely polluted mining areas (e.g., [16]). Generally, the relationship between Hg bioaccessibility in soil and the plant's ability to take up this element is low compared to other risk elements, such as cadmium [1].

Phytotoxicity symptoms have been reported in Hg-polluted soils. As reported by Natasha et al. [17], Hg exerts toxic effects in plants even at low concentrations, resulting in growth retardation, photosynthesis inhibition, reactive oxygen species generation, lipid membrane oxidation, and DNA and protein damage. Furthermore, due to Hg's effects on life, it is highly imperative to monitor and control the biogeochemical behavior of Hg in soil–plant–human systems [17]. For instance, soil Hg levels ranging from 1 to 50 mg/kg result in negative effects on seed germination and root elongation of *Lepidium sativum* and *Solanum lycopersicum* under model conditions [18]. Moreno-Jiménez et al. [19] tested the ability of two plant species, *Rumex induratus* and *Marrubium vulgare*, to take up Hg under model conditions in a soilless culture. Both plant species originated from a location extremely polluted by Hg due to long-term mining activity. They observed a high accumulation ability and phytotoxicity symptoms (such as reduced plant growth, chlorophyll concentration, and nutrient translocation) with increasing Hg levels in the medium, but *R. induratus* showed better tolerance to Hg pollution than *M. vulgare*, as determined by the oxidative stress indices. Moreover, Moreno-Jiménez et al. [20] confirmed enhanced tolerance to extremely high Hg concentrations in *R. induratus* naturally growing in a Hg-polluted area compared to plants of the same species originating from a non-Hg polluted site. The higher tolerance of the plants originating from the mining area was documented, for instance, by lower Hg uptake and enhanced thiol synthesis by plants from the polluted site compared to plants from the unpolluted area. These findings clearly document the importance of plant species in Hg uptake and tolerance. Therefore, a Hg-tolerant plant community can be expected in long-term abandoned Hg ore mining areas.

In the case of Hg, no hyperaccumulating plant has yet been discovered [21]. However, the extensive investigation of the Hg bioaccumulation ability of various plant species has revealed promising results. Duan et al. [22] reviewed the available plant species suitable for phytoremediation (i.e., phytoextraction) of Hg-polluted soil, where the analysis of the available data identified *Cyrtomium macrophyllum* and *Phragmites australis* as promising Hg accumulators. A similar review by Li and Tse [23] mentioned that *Achillea millefolium* accumulated high amounts of Hg. Among the other potential Hg-accumulating plants, *Brassica juncea* is frequently mentioned [24]. Although only confirmed in soilless culture, Alcantara et al. [25] showed the promising Hg accumulation ability of *Manihot esculenta*. García-Mercado et al. [26] demonstrated the efficient ability of *Polypogon monspeliensis* and *Cyperus odoratus* plants to remove Hg from Hg-polluted mine wastes in a model pot experiment. Among aquatic macrophytes, *Heliconia psittacorum* showed promising ability to accumulate Hg in areas with artisanal mining activity [27].

There are several abandoned cinnabar mining areas in the Czech Republic, but information concerning Hg pollution levels in soils in the vicinity of mine shafts and waste dumpsites, and the potential environmental risks of these Hg levels, is limited. For this investigation, two historical cinnabar mining areas were chosen: (i) the Hg mine in Horní Luby, situated in North Bohemia (Czech Republic), and (ii) Jedová Hora in Central Bohemia (Czech Republic). In the Horní Luby mine, the ore consists of pure cinnabar [28]. The main mining phase occurred from 1520 to 1540, followed by a second peak from 1560 to 1570, when Hg production was 6–15 tonnes per annum. At that time, the mine was listed among the largest Hg producers in Europe. The mine was finally abandoned in 1597, and several unsuccessful attempts to restore mining activity from the 17th to the 19th century were not very extensive [29]. In the Jedová Hora mine, where cinnabar was mined as a by-product of Ordovician iron ore (mostly containing haematite and siderite), mining continued from the 18th century until the end of the 19th century. The estimated Hg production was several tonnes per annum [30]. At present, both locations are covered by mixed (mostly deciduous) forests, with a dense cover of herbaceous plants.

Since no relevant information concerning Hg contents in soils is available for the Horní Luby location, data on soil Hg contents, Hg fractions and compounds, as well as the proportions of organomercurials and gaseous elemental mercury ($GEM_{soil\text{-}air}$), were only determined for the Jedová Hora location [31,32]. These investigations showed elevated Hg contents in the soil profiles, where approximately 50%–80% of total Hg in all soil horizons was present as relatively insoluble HgS. The remaining 20%–50% of total Hg was related to Hg adsorbed on mineral surfaces, Fe-oxyhydroxides, or clay minerals. However, the potential soil–plant transport of Hg has not yet been studied at both locations. At the Horní Luby location, elevated Hg contents can also be expected, and the potential adverse effects of Hg on the biota cannot be excluded. Therefore, almost no reliable information is available concerning the potential environmental risk of elevated Hg contents in soil at both locations. Therefore, this study focuses on the assessment of plants' ability to take up Hg from polluted soil as affected by location, plant family, and species.

## 2. Materials and Methods

### 2.1. Sampling and Sample Processing, Determination of $GEM_{soil\text{-}air}$

Soil samples were collected in the vicinity of important points connected to mining activity, such as outfalls of abandoned drifts and mine waste heaps. At each location, there were 7 sampling points, and the soil samples were taken from a depth of 0–20 cm, with 5 soil samples taken from an area of ca. 5 $m^2$ at the individual sampling points. The sampling grid differed at individual locations because of the different distribution of the mining residues at these locations. The most important characteristics of the sampling points, including the GPS coordinates, are presented in Supplementary Table S1. A detailed description of the whole mining area is provided elsewhere [28–31].

A Lumex RA-915$^{+}$ instrument (Lumex Ltd., St. Petersburg, Russia) was used for $GEM_{soil\text{-}air}$ measurement of the soil–air interface at five sampling points at the Horní Luby location. The mercury LOD is as low as 2 ng $m^{-3}$, with a method bias of ±20%. To obtain comparable results, all $GEM_{soil\text{-}air}$ measurements were performed using the same procedure tested and described previously, especially at the Jedová Hora location [32,33]. GEM concentrations were measured inside a PE chamber, which was placed tightly onto the uncovered surface area. After 4 min of equilibration, the mouth of an inlet tube of the analyzer was placed at a height of 5 cm above the surface and a maximum GEM value was recorded. Subsequently, soil samples from the chamber's position were collected and processed in the same manner for the sampling area monitoring.

At each sampling point, representative samples of dicotyledonous plants were collected at the flowering stage, with aboveground biomass and roots separated. At the Jedová Hora location, 54 plant species were found representing the Fabaceae, Violaceae, Rosaceae, Rubiaceae, Plantaginaceae, Asteraceae, Urticaceae, Euphorbiaceae, Hypericaceae, and Lamiaceae families. At the Horní Luby location, 36 plant species were collected, representing the Rubiaceae, Apiaceae, Oxalidaceae, Lamiaceae, Asteraceae, Hypericaceae, Urticaceae, Rosaceae, Violaceae, and Ranunculaceae families. Moreover, the biomasses of the fern *Dryopteris filix-mas* (Dryopteridae) and monocotyledonous *Convallaria majalis* (Convallariaceae) were sampled at this location as the dominant plant species. All plant species are listed in Supplementary Tables S2 and S3.

### 2.2. Determination of the Total Hg Content in Soils and Plants

Soil samples were freeze-dried (FreeZone® 6 Liter Benchtop Freeze Dry System; Labconco Corp. Kansas City, MO, USA), ground in a mortar, and passed through a 2 mm plastic sieve before analysis. The plant biomass was gently washed in deionized water using an ultrasonic bath, freeze-dried as mentioned above, and homogenized (IKA-Werke GmbH & Co. KG, Staufen, Germany) before analysis. To determine the total Hg content in soils and plants, an AMA-254 single-purpose atomic absorption spectrometer (LECO model, Altec, Czech Republic) was used. In this device, the samples were dried and subsequently burned in a stream of oxygen in a closed system. Subsequently, released Hg vapor was trapped on a gold amalgamator and then quickly released as $Hg^0$ by heating. The released Hg vapor was measured with an atomic absorption spectrometer at a wavelength of 253.65 nm. A low-pressure Hg discharge lamp was used as a source of radiation in this case [34]. All measurements were conducted in triplicate, with an LOD of 0.01 ng Hg. Standard reference material SRM NIST 1573a was applied for quality assurance of the data, where the certified value was 0.034 ± 0.004 mg/kg of Hg, and the determined value was 0.032 ± 0.005 mg/kg of Hg.

### 2.3. Estimation of Hg Accumulation Ability by Plants

The bioaccumulation factor (*BAF*) was used to describe Hg uptake in the investigated plants, as follows:

$$BAF = \frac{C_{plant}}{C_{soil}} \quad (1)$$

where $C_{plant}$ is the Hg concentration in plants (aboveground biomass, BAFs, or roots, BAFr) and $C_{soil}$ is the total Hg concentration in the soil. A factor level $\geq 1$ indicates an enhanced ability of plants to take up Hg from the soil.

The translocation factor (*TF*) was used to describe the ability of investigated plants to translocate Hg from roots to the aboveground biomass, as follows:

$$TF = \frac{C_{shoots}}{C_{roots}} \quad (2)$$

where $C_{shoots}$ is the Hg concentration in aboveground biomass, and $C_{roots}$ is the total Hg concentration in the plant roots. A factor level $\geq 1$ indicates an enhanced ability of plants to translocate Hg into the aboveground biomass [35].

### 2.4. Statistics

Statistical evaluation of the analytical datasets for Hg contents in plants was performed using Statistica 12.0 software (www.StatSoft.com; accessed on 15 July 2024). Descriptive statistics was used to describe the BAFs and TFs datasets, where minimum, maximum, average, standard deviation, median, and median of absolute deviations (MAD) were calculated. The individual datasets did not match the parameters of a normal data distribution using Shapiro–Wilk W normality and Levene variance tests; therefore, the non-parametric Kruskal–Wallis test was used to assess statistically significant differences in measured parameters among plant families. To compare the measured parameters between the two sampling locations, the Mann–Whitney test was applied. Spearman's rank correlation coefficients (ρ) were applied to determine the potential interrelationships among the experimental factors, where both the whole dataset and datasets for the individual locations were compared. In all cases, $p < 0.05$ was considered significant.

## 3. Results and Discussion

### 3.1. Total Hg Content in Soils at the Hg-Polluted Areas

The total Hg content in the soil varied between 0.221 and 135 mg/kg at Jedová Hora and between 3.34 and 117 mg/kg at Horní Luby (Supplementary Table S1). Public Notice No. 153/2016 [36] in the Czech Republic focused on the conditions for the protection of agricultural soils, presenting a preventive value (i.e., the element content indicating the elevated level for the total Hg content in soil without a harmful effect on the soil and biota) of 0.3 mg/kg. Except at one sampling point, all soil Hg values exceeded this limit, i.e., 93% of the total sample number. The indicative value at which the soil Hg content represents a potential risk for crop contamination is 1.5 mg/kg of Hg, and the level that directly threatens human and animal health is 20 mg/kg of Hg [36]. Most of the soil samples substantially exceeded these limits. As reviewed by Natasha et al. [17], an extremely high soil Hg content can occur in Hg mining areas, such as 1709 mg/kg found in Asturias, Spain, or 450 mg/kg in Guizhou, China. In this context, the maximum Hg values at both Jedová Hora and Horní Luby can be considered relatively low compared to other mining areas but far from the typical soil Hg values determined at various unpolluted locations in the Czech Republic, varying from 0.02 to 0.2 mg/kg [37].

### 3.2. $GEM_{soil\text{-}air}$ Concentrations at Hg-Polluted Areas

The soils at the Jedová Hora location are characterized by low Hg mobility; relatively insoluble HgS has been identified as a dominant Hg compound [31]. Additionally, Sysalová et al. [32] found minor proportions of elemental mercury ($Hg^0$), methylmercury ($MeHg^+$), and phenylmercury ($PhHg^+$) at the Jedová Hora location. At Horní Luby, the former cinnabar mining area, the dominance of HgS as well as the limited mobility of Hg was also expected. For an assessment of the potential impact of soil Hg pollution on the surrounding environment, the proportions of $GEM_{soil\text{-}air}$ can provide reasonable information. At the Jedová Hora location, the previously determined maximum $GEM_{soil\text{-}air}$ concentrations [32] varied between 5 and 15 ng/m$^3$ in the soils, with a total Hg content in the range of 11–144 mg/kg. Therefore, because of the expected similarity of the mineralogical composition of both mining areas, comparable $GEM_{soil\text{-}air}$ concentrations are also expected. In this study, the $GEM_{soil\text{-}air}$ concentrations were determined at 5 sampling points at the Horní Luby location, where at sampling point No. 6 (one of the mine waste dumpsites), with a total soil Hg content of 3.81 mg/kg, the measured $GEM_{soil\text{-}air}$ concentrations did not differ from the levels measured outside the measuring chamber.

At sampling points 1 (measured next to the entrance to the shaft), 2, and 5 (mine waste dumpsites), where the total soil Hg content varied from 39.2 to 56.1 mg/kg, the maximum $GEM_{soil-air}$ concentration varied between 2 and 9 $ng/m^3$. Exceptionally high maximum $GEM_{soil-air}$ values reaching up to 498 $ng/m^3$ in the soil with a total Hg content of 135 mg/kg were found at sampling point No. 4 (the bank of the lake lying among the dumpsites). Therefore, the results indicated a lower $GEM_{soil-air}$ concentration compared to the Jedová Hora location but with the occurrence of hotspots, which deserves more detailed investigation in the future. Similarly, Fernández-Martínez et al. [38] recorded low free Hg(0) contents in soils ranging from 2.1 to 113 mg/kg, with percentages varying between 6 and 20% of the total Hg content. In this case, however, the pyrolysis technique was used to determine free Hg(0); therefore, the Hg proportions are not comparable to the $GEM_{soil-air}$ concentrations determined in this study. Generally, the $GEM_{soil-air}$ concentrations at both locations cannot be underestimated, and their impact on the Hg content in biota should be elucidated in more detail.

### 3.3. Total Hg Content in Plants in the Hg-Polluted Areas

A detailed list of the Hg content in all analyzed plants is presented in Supplementary Tables S2 and S3. The total Hg contents in the roots and shoots of plants regardless of the plant species and sampling point are presented in Figure 1. The results of the statistical evaluation of the data showed significantly ($p < 0.05$) higher Hg levels in roots at the Horní Luby location compared to Jedová Hora, whereas no significant differences were observed between locations for the shoot Hg content. The individual locations are characterized by different plant communities, where, in most cases, different plant species were found within one family at both locations (for instance, *Galium odoratum* at Jedová Hora vs. *G. aparine* at the Horní Luby location). Moreover, the composition of plant species differed even among the individual sampling points, with the exception of *Viola sylvestris* found at all seven sampling points at Jedová Hora. In this plant, the Hg contents in roots varied in the range from 0.179 to 9.78 mg/kg (median value 0.309 mg/kg) in roots, and from 0.011 to 0.590 mg/kg (median value 0.075 mg/kg) in shoots. These data indicate that the Hg accumulation ability of this plant did not differ from other species; the maximum value was reached at the sampling point with the highest Hg content (135 $\pm$ 1 mg/kg). The results confirmed the ability of plants to accumulate Hg predominantly in the roots compared to the aboveground biomass in most plant species [39]. The median Hg content in plants at the Jedová Hora location was 0.071 mg/kg in the aboveground biomass and 0.331 mg/kg in the roots. The maximum measured Hg values were found for both aboveground biomass and roots of *Onobrychis viciifolia* (Fabaceae) (0.829 and 13.4 mg/kg, respectively). At the Horní Luby location, the median Hg content in plants was 0.11 mg/kg in the aboveground biomass and 1.01 mg/kg in the roots. The maximum measured Hg value was 0.575 mg/kg for the aboveground biomass of *Ficaria verna* and 14.8 mg/kg in the roots of *Galium aparine*.

Fernández et al. [40] found a high Hg content in some plant species in the old mining area in Cantabrian Zone of the Iberian Massif, Spain; for instance, they found 84 mg/kg of Hg in shoots of *Festuca rubra*, 78 mg/kg in *Leontodon taraxacoides*, and 77 mg/kg in *Equisetum telmateia*. However, the total Hg content in the soils in this case substantially exceeded those recorded in this study, reaching up to 7963 mg/kg. Similarly, Fernández-Martínez et al. [38] investigated Hg uptake by plants growing at old cinnabar mining sites in Spain (total Hg content in soils ranged from 36 to 1709 mg/kg). The Hg content in the studied plants ranged from 0.6 to 48 mg/kg in the aboveground biomass and from 1.8 to 160 mg/kg in the roots, where the plant species with the highest amount of total Hg was *Crupina vulgaris* (Asteraceae). Both studies confirmed a wide range of plant Hg contents as affected by plant species and soil Hg levels. Raj [41] tested the Hg accumulation ability of *Brassica juncea* plants in a model pot experiment, in which the plants were grown in soil polluted by coal mining activities. In this case, higher Hg accumulation was confirmed in roots compared to shoots. Millan et al. [42] investigated the potential Hg accumulation ability of riparian plant species growing along the river running through a former Hg mining area. These authors compared the Hg content of aerial parts of *Flueggea tinctoria*, *Tamarix canariensis*, *Nerium oleander*, *Typha domingensis*, and *Phragmites australis* and found that the Hg content differed according to species. However, the low Hg mobility in soils and sediments resulted in a low Hg content in the aerial parts of plants, similar to this study.

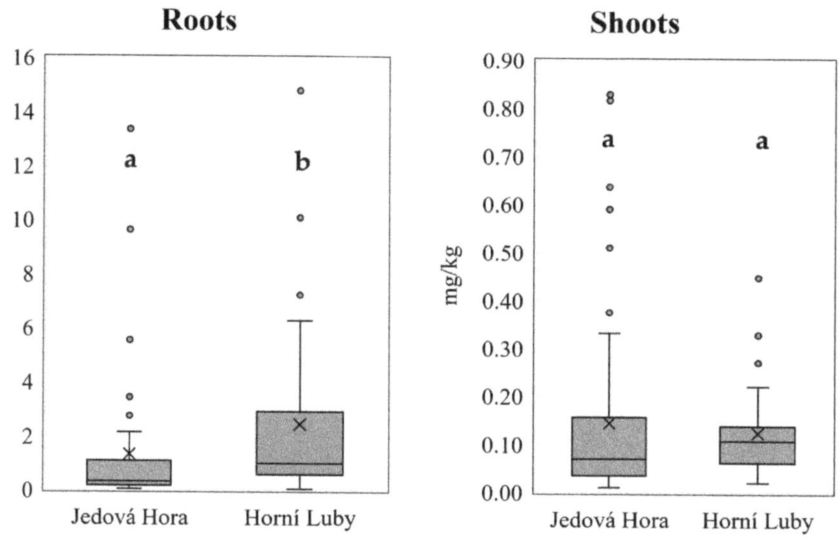


**Figure 1.** The total Hg contents in the roots and shoots of plants at both locations, regardless of plant species and family; n = 54 for Jedová Hora, and n = 36 for the Horní Luby location. Horizontal lines drawn inside the boxes indicate the median, x indicates the average, and the points outside the whisker boundaries are outliers. The boxplots marked by the same letter do not significantly differ at $p < 0.05$.

The differences in Hg accumulation in roots and shoots among the plant families are presented in Figures 2 and 3. The results indicate an enhanced Hg accumulation ability of Fabaceae plants at Jedová Hora (Figure 2); however, this trend was not statistically supported; the differences among the plant families at this location were not significant at $p < 0.05$. Among the Fabaceae family species, Esteban et al. [43] confirmed that *Lupinus albus* was tolerant to increased Hg levels in the nutrient solution. Carrasco-Gil et al. [44] studied the change in biomass growth of alfalfa (*Medicago sativa*), also belonging to the Fabaceae family, when grown in containers of agricultural soil obtained from a Hg-contaminated site (Almadén, Spain). Some of the samples were grown with fertilizers (NPK) and some without added fertilizers. The results indicate that an increased N supply during growth prevents oxidative stress in roots and improves root development and, thus, Hg uptake from the soil.

The positive effect of N on plant tolerance to Hg indicates the potential role of N-fixing bacteria. Nonnoi et al. [45] confirmed that the increased tolerance of plants from the Fabaceae family to Hg was related to their interaction with microorganisms in the soil (especially the genus Rhizobium), which are capable of binding N and affecting the bioavailability of hazardous metals. These authors identified rhizobial strains of bacteria in plants from the Fabaceae family (*Medicago* spp. and *Trifolium* spp.) that are highly tolerant to Hg. The mechanisms of Hg tolerance in Fabaceae plants have already been investigated. Recently, Raghupathy and Arunachalam [46] showed that Hg-tolerant rhizobia were responsible for enhanced plant tolerance and that rhizobium alleviated Hg phytotoxicity in plants via the reduction of Hg to volatile $Hg^0$ form and subsequent tight Hg sequestration in the root cell wall and limited translocation of Hg to the aboveground biomass. They also reported increased free radical scavenging in nodules exposed to elevated Hg levels.

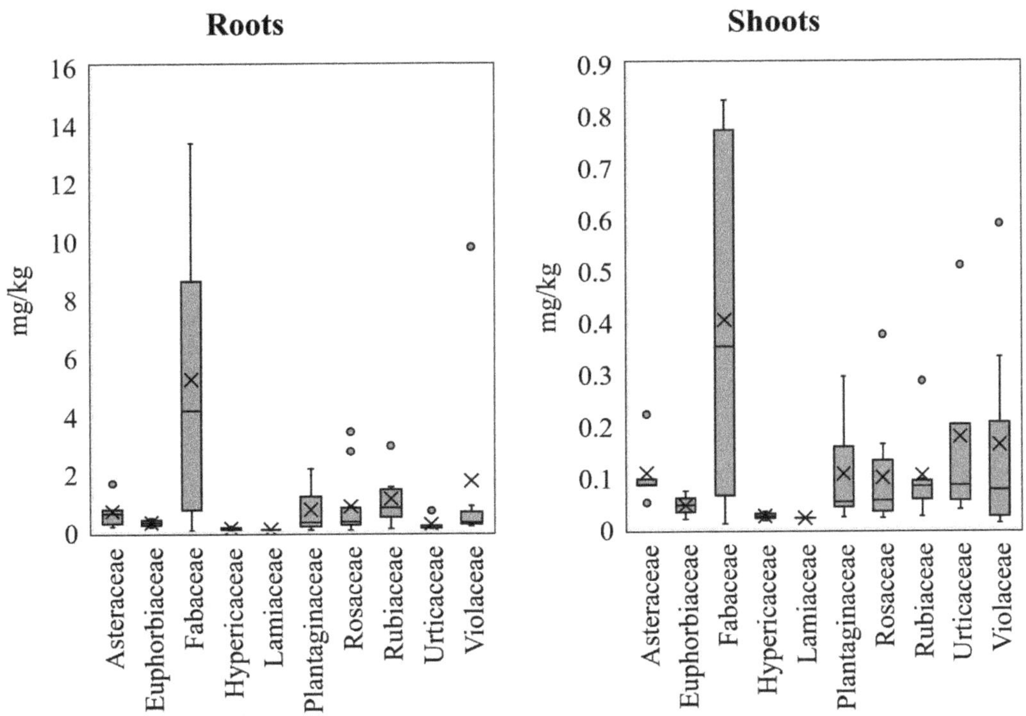


**Figure 2.** The total Hg contents in the roots and shoots of plants at the Jedová Hora location according to the individual plant families; n = 54. Horizontal lines drawn inside the boxes indicate the median, x indicates the average, and the points outside the whisker boundaries are outliers.

Numerous scientists consider Fabaceae plants suitable species for potential phytoremediation of Hg-polluted areas because of their tolerance to elevated Hg contents and relatively good ability to accumulate Hg [47,48]. Moreover, although the ability of these plants to accumulate Hg is far from that of hyperaccumulation, they have shown promising characteristics for potential phytoremediation, such as their ability to grow quickly, produce high biomass yields, are be easy to harvest [48], and the fact that they mostly accumulate Hg in belowground organs, posing a lesser threat of contamination and bioaccumulation across the food chain. Additionally, attempts to obtain a suitable cultivar to genetically improve the phytoremediation results have already been made, for example, by de la Torre et al. [48] with *Medicago truncatula* plants. However, the fate of Fabaceae plants in the polluted soil, the mechanisms of their tolerance to the high Hg content and Hg accumulation ability, the role of soil properties, and the transformation of Hg species in both plants and soil need to be elucidated in further research.

At Horní Luby (Figure 3), no species representing the Fabaceae family were found. Although Ranunculaceae and Rosaceae plants showed an enhanced total Hg content in shoots, the differences were not significant ($p < 0.05$). Su et al. [49] found that *Pteris vittata* showed low stress symptoms resulting from Hg exposure and had the highest Hg accumulation, thus recommending *P. vittata* as a potential candidate for Hg phytoextraction from polluted soil. The dominant fern species growing at this location, *Dryopteris filix-mas*, was analyzed, showing that the Hg content in the aboveground biomass varied from 0.055 to 0.138 mg/kg and in roots from 0.48 to 1.01 mg/kg. Thus, the Hg values in this species were close to the median values for this location. Although the Hg content in *Dryopteris filix-mas* and *D. affinis* as determined by Fernández-Martínez et al. [38] exceeded those found in this study (most probably due to lower soil Hg content compared to the cited paper), the Hg content was significantly lower in fern species than in dicotyledonous and monocotyledonous herbaceous plants analyzed by these authors. Similarly, Wang et al. [50] did not show elevated Hg levels in *Dryopteris setosa* plants. Therefore, the results indicate the low ability of *Dryopteris* sp. to accumulate Hg.

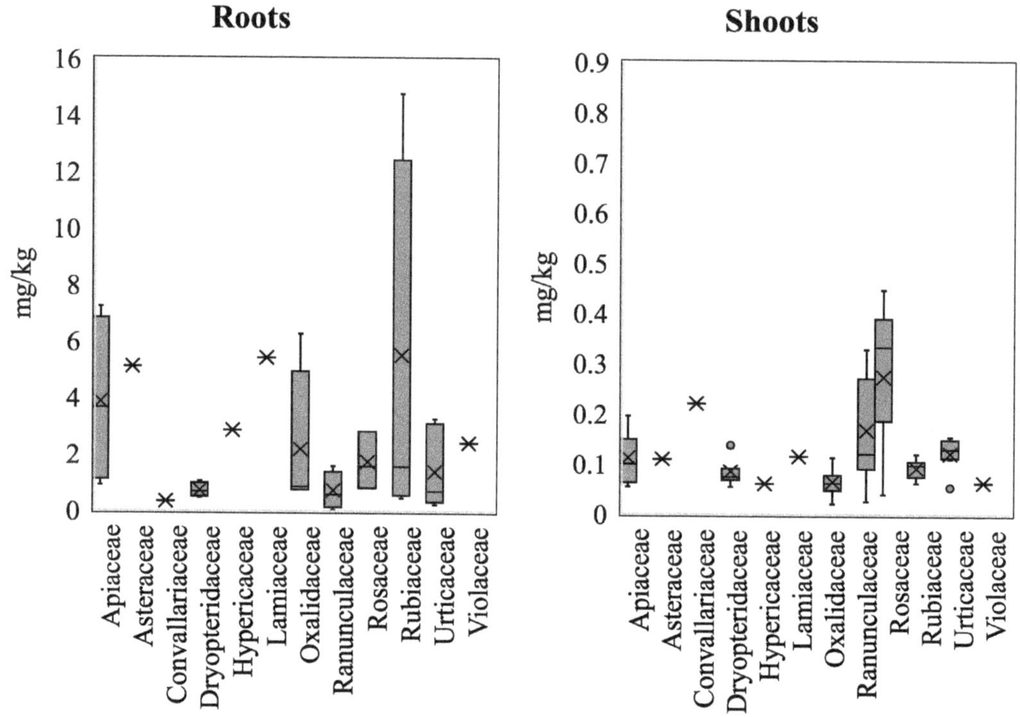


**Figure 3.** The total Hg contents in roots and shoots of plants at the Horní Luby location according to the individual plant families; n = 36. Horizontal lines drawn inside the boxes indicate the median, x indicates the average, and the points outside the whisker boundaries are outliers.

### *3.4. Ability of Plant Biomass to Accumulate and Translocate Hg from Polluted Soils*

The BAF and TF values for the individual locations are summarized in Figures 4 and 5. The Mann–Whitney test showed no significant ($p < 0.05$) differences between BAF values for plant roots, whereas significantly ($p < 0.05$) higher BAFs were recorded for shoots at Jedová Hora compared to Horní Luby. Similarly, translocation factors (TFs), indicating the plant's ability to translocate Hg from the shoots to the roots, were significantly ($p < 0.05$) higher at Jedová Hora. The $BAF_{shoots}$, $BAF_{roots}$, and TFs for the individual families are presented in Tables 1 and 2. At Jedová Hora and Horní Luby, the BAFs for the roots and shoots, as well as the TFs, did not significantly ($p < 0.05$) differ among the plant families.

Considering all plant samples, both root and shoot Hg contents were significantly ($p < 0.05$) correlated with the Hg content in the soil (Supplementary Table S4). The Spearman's rank correlation coefficients were $\rho = 0.51$ for roots and $\rho = 0.34$ for shoots, indicating soil–plant transport as the major pathway of Hg uptake by plants. Similarly, Millán et al. [42] found positive correlations between the Hg contents in riparian plants and in soils and sediments along the river in a Hg-polluted area. However, a different pattern was identified when the locations were assessed separately. The relationships between soil and plant Hg contents at Jedová Hora were similar as in the case of the whole dataset ($\rho = 0.46$ for roots and $\rho = 0.38$ for shoots; significant at $p < 0.05$), but no significant relationships were recorded between plant and soil Hg contents for roots or shoots at Horní Luby.

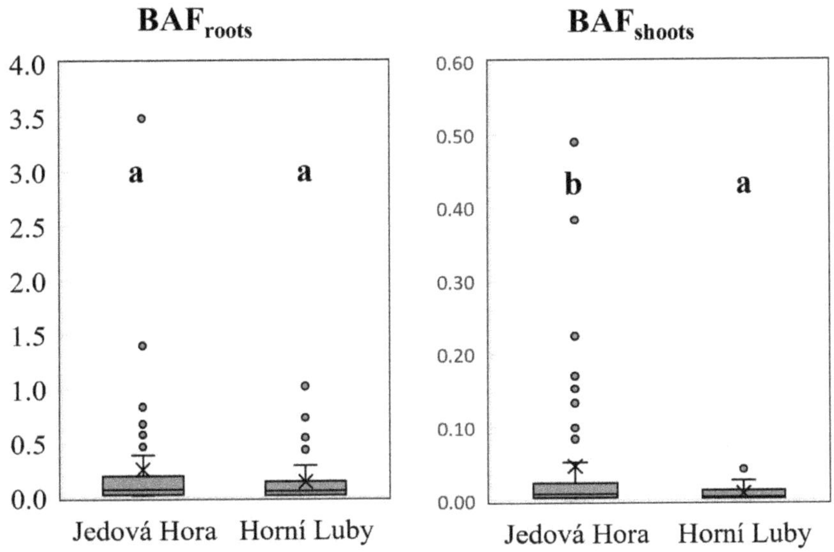


**Figure 4.** The BAF values of plants at both locations, regardless of the plant species and family; n = 54 for Jedová Hora, and n = 36 for the Horní Luby location. Horizontal lines drawn inside the boxes indicate the median, x indicates the average, and the points outside the whisker boundaries are outliers. The boxplots marked by the same letter did not significantly differ at $p < 0.05$.

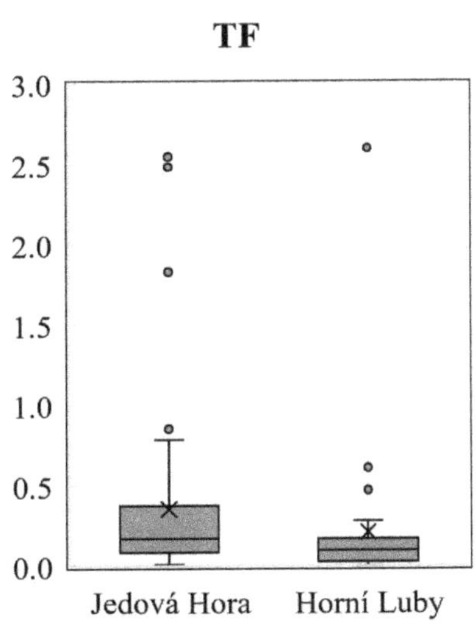


**Figure 5.** The TF values of plants at both locations, regardless of the plant species and family; n = 54 for Jedová Hora, and n = 36 for the Horní Luby location. Horizontal lines drawn inside the boxes indicate the median, x indicates the average, and the points outside the whisker boundaries are outliers. The boxplots marked by the same letter did not significantly differ at $p < 0.05$.

**Table 1.** Descriptive statistics of the Hg accumulation ability of plants at the Jedová Hora location, categorized by plant families.

| | Asteraceae | Fabaceae | Plantaginaceae | Rosaceae | Rubiaceae | Urticaceae | Violaceae |
|---|---|---|---|---|---|---|---|
| | | | | $BAF_{roots}$ | | | |
| n | 5 | 6 | 9 | 11 | 6 | 5 | 7 |
| Min | 0.033 | 0.015 | 0.007 | 0.004 | 0.004 | 0.001 | 0.020 |
| Max | 3.49 | 0.675 | 1.44 | 0.386 | 0.467 | 0.080 | 1.40 |
| Average | 0.903 | 0.348 | 0.248 | 0.082 | 0.186 | 0.022 | 0.241 |
| Standard deviation | 1.486 | 0.271 | 0.456 | 0.116 | 0.187 | 0.032 | 0.510 |
| Median | 0.084 | 0.360 | 0.109 | 0.033 | 0.116 | 0.011 | 0.044 |
| MAD | 0.560 | 0.240 | 0.089 | 0.025 | 0.086 | 0.008 | 0.024 |
| | | | | $BAF_{shoots}$ | | | |
| n | 5 | 6 | 9 | 11 | 6 | 5 | 7 |
| Min | 0.005 | 0.003 | 0.002 | 0.001 | 0.001 | 0.004 | 0.003 |
| Max | 0.38 | 0.13 | 0.17 | 0.49 | 0.22 | 0.01 | 0.051 |
| Average | 0.131 | 0.050 | 0.042 | 0.049 | 0.044 | 0.007 | 0.015 |
| Standard deviation | 0.170 | 0.054 | 0.058 | 0.146 | 0.089 | 0.003 | 0.017 |
| Median | 0.025 | 0.029 | 0.015 | 0.003 | 0.011 | 0.007 | 0.009 |
| MAD | 0.021 | 0.025 | 0.010 | 0.002 | 0.006 | 0.003 | 0.006 |
| | | | | TF | | | |
| n | 5 | 6 | 9 | 11 | 6 | 5 | 7 |
| Min | 0.058 | 0.007 | 0.020 | 0.023 | 0.015 | 0.119 | 0.036 |
| Max | 0.35 | 0.85 | 0.60 | 2.49 | 0.48 | 2.55 | 0.420 |
| Average | 0.218 | 0.250 | 0.268 | 0.445 | 0.185 | 1.09 | 0.181 |
| Standard deviation | 0.127 | 0.308 | 0.193 | 0.727 | 0.170 | 1.05 | 0.157 |
| Median | 0.276 | 0.172 | 0.286 | 0.136 | 0.168 | 0.601 | 0.108 |
| MAD | 0.251 | 0.084 | 0.156 | 0.108 | 0.106 | 0.368 | 0.071 |

MAD, median of absolute deviations; families with $n \geq 3$ were considered for evaluation.

Concerning the Hg accumulation ability estimated as BAF and TF values at the individual locations, significant ($p < 0.05$) Spearman's rank correlation coefficients showed decreasing BAFs and TFs with increasing soil Hg content ($\rho = -0.50$ for $BAF_{roots}$, $\rho = -0.76$ for $BAF_{shoots}$, and $\rho = -0.30$ for TF). Considering the individual locations separately, significant ($p < 0.05$) relationships were only confirmed for BAFs, with $\rho = -0.55$ for $BAF_{roots}$ and $\rho = -0.73$ for $BAF_{shoots}$ at Jedová Hora and $\rho = -0.64$ for $BAF_{roots}$ and $\rho = -0.94$ for $BAF_{shoots}$ at Horní Luby. These results indicate the plants' ability to limit Hg uptake in highly polluted soils to prevent the potential phytotoxic effects of elevated Hg levels in plants.

In a model pot experiment, Yang et al. [51] evaluated Hg accumulation in the edible parts of various leafy vegetables, including *Spinacia oleracea*, *Allium porrum*, *Foeniculum vulgare*, *Coriandrum sativum*, *Brassica rapa chinensis*, *Cichorium intybus*, and *Lactuca sativa* and determined the soil Hg levels. Although Hg was added into the soil artificially, therefore resulting in higher Hg mobility compared to the mining areas, the BAF values were far less than one in all cases, confirming the generally low plant availability of Hg. Therefore, these authors reported a similarly low plant ability to take up Hg, as in this study. Comparing the Hg accumulation ability of plants with other risk elements, it is among the lowest. Peralta-Videa et al. [52] found the following order for plant bioaccumulation ability of risk elements: Cd > Zn > Cu > Pb > Hg. The low plant availability of Hg compared to other elements, especially Cd, was confirmed in our previous investigation [53], in which plants growing in the Hg-polluted area of an abandoned gold mine were sampled.

**Table 2.** Descriptive statistics of the Hg accumulation ability of plants at the Horní Luby location categorized by plant families.

| | Apiaceae | Dryopteridaceae | Oxalidaceae | Ranunculaceae | Rosaceae | Rubiaceae | Urticaceae |
|---|---|---|---|---|---|---|---|
| | | | | $BAF_{roots}$ | | | |
| n | 4 | 4 | 4 | 5 | 3 | 5 | 6 |
| Min | 0.027 | 0.014 | 0.013 | 0.005 | 0.015 | 0.006 | 0.006 |
| Max | 0.73 | 0.15 | 0.29 | 0.06 | 0.16 | 1.02 | 0.21 |
| Average | 0.245 | 0.076 | 0.102 | 0.026 | 0.077 | 0.325 | 0.077 |
| Standard deviation | 0.327 | 0.061 | 0.127 | 0.024 | 0.075 | 0.424 | 0.075 |
| Median | 0.111 | 0.072 | 0.052 | 0.011 | 0.056 | 0.138 | 0.065 |
| MAD | 0.042 | 0.046 | 0.020 | 0.006 | 0.042 | 0.132 | 0.042 |
| | | | | $BAF_{shoots}$ | | | |
| n | 4 | 4 | 4 | 5 | 3 | 5 | 6 |
| Min | 0.001 | 0.001 | 0.0002 | 0.0002 | 0.001 | 0.001 | 0.001 |
| Max | 0.013 | 0.022 | 0.017 | 0.028 | 0.045 | 0.019 | 0.042 |
| Average | 0.006 | 0.010 | 0.006 | 0.013 | 0.018 | 0.007 | 0.012 |
| Standard deviation | 0.005 | 0.009 | 0.008 | 0.013 | 0.024 | 0.008 | 0.016 |
| Median | 0.005 | 0.009 | 0.003 | 0.010 | 0.007 | 0.003 | 0.007 |
| MAD | 0.002 | 0.006 | 0.002 | 0.009 | 0.006 | 0.002 | 0.005 |
| | | | | TF | | | |
| n | 4 | 4 | 4 | 5 | 3 | 5 | 6 |
| Min | 0.010 | 0.071 | 0.003 | 0.022 | 0.049 | 0.007 | 0.036 |
| Max | 0.21 | 0.15 | 0.15 | 2.60 | 0.28 | 0.14 | 0.61 |
| Average | 0.070 | 0.120 | 0.075 | 0.758 | 0.150 | 0.067 | 0.213 |
| Standard deviation | 0.097 | 0.035 | 0.062 | 1.048 | 0.121 | 0.058 | 0.214 |
| Median | 0.028 | 0.129 | 0.073 | 0.471 | 0.118 | 0.068 | 0.155 |
| MAD | 0.014 | 0.012 | 0.042 | 0.264 | 0.069 | 0.056 | 0.113 |

MAD, median of absolute deviations; families with $n \geq 3$ were considered for evaluation.

Egler et al. [54] found that wildlife plant species growing in Hg-polluted gold mining areas predominantly belonged to excluders with limited Hg translocation into the aboveground biomass. In accordance with these findings, Fernández et al. [40] identified *Lotus corniculatus* as an excluder plant able to grow without any problems at Hg-polluted sites. In contrast, Umlaufová et al. [53] recorded the highest BAF (BAF = 0.37) for the aboveground biomass of *L. corniculatus* among the analyzed plant species growing in Hg-polluted soil, indicating that all plant species (not exceeding BAF = 1) were found here. The low BAF values were fully confirmed at the Horní Luby location (Table 2), where a BAF up to 1 was only recorded for roots of *Galium aparine* (Rubiaceae). At the Jedová Hora location (Table 1), several plant species from the families Asteraceae, Plantaginaceae, and Violaceae showed elevated BAF values for roots with the highest $BAF_{roots}$ for *Achillea millefolium* (Asteraceae). However, this value was achieved at sampling point 5, characterized by the lowest total Hg content (Supplementary Table S1). This result should support the statement that the plant's ability to accumulate Hg decreases with increasing soil Hg content; unfortunately, this species occurred solely at this sampling point, making any comparison with other sampling points impossible.

The predominant accumulation of Hg in roots compared to the aboveground biomass was confirmed. Sasmaz et al. [55] monitored Hg uptake by plants growing naturally in an extremely polluted mining area, where BAF values calculated separately for roots and shoots were also far less than one. However, TFs documenting the ability of plants to translocate Hg from roots to the aboveground biomass frequently exceeded one (for instance, in the case of *Anchusa arvensis*, *Centaurea cyanus*, *Onosma* sp., *Phlomis* sp., *Silene compacta*, and *Verbascum thapsus*), indicating potential Hg accumulation in the shoots of these species. All of these findings are in accordance with this study, showing that the Hg content in plants is low compared to the soil Hg content. Concerning TFs, a relatively high ability of plants to translocate Hg from roots to aboveground biomass was recorded for the families Ranunculaceae, Urticaceae, and Rosaceae at both locations (Tables 1 and 2), indicating the phytoremediation potential of these plants.

To evaluate the Hg pathway from the polluted soil to plants, it is necessary to consider the volatile proportions of Hg. The significant release of Hg and its compounds from the soil to the atmosphere is well documented, including the role of soil microorganisms [56]. As already discussed, detectable $GEM_{soil\text{-}air}$ concentrations were recorded at both Jedová Hora and Horní Luby. Meloni et al. [57] compared Hg uptake by various plant species (*Castanea sativa, Sambucus nigra, Verbascum thapsus, Populus* spp., *Salix* spp., *Acer pseudoplatanus*, and *Robinia pseudoacacia*) growing in Hg-polluted soils from various mining areas (both active and abandoned). In this case, leaves were considered the main pathway for Hg uptake, but BAF values remained below one. In contrast, the results of this study did not clearly show the pathway of Hg uptake but indicated the importance of the soil–plant pathway. Among the factors affecting these differences, the different plant communities analyzed (trees vs. herbaceous plants) may play a role.

García-Sánchez et al. [58] investigated the soil–plant transport of Hg in the area of a former cinnabar mine, where the mobile Hg contents in soil were low compared to the total content. However, the Hg content in plants was relatively high, reaching 37.6 mg/kg in *Marrubium vulgare*. Similar to Gworek et al. [1], these authors suggested that the plants predominantly take up atmospheric Hg, where leaf-specific surfaces are an important factor in Hg uptake. In this context, plants with small to medium size and/or with medium to highly corrugated leaves, such as *M. vulgare*, *Bromus madritensis*, and *Trifolium angustifolium*, showed the highest ability to accumulate Hg. In contrast, Hg uptake by roots can reflect the Hg content in the soil [59]. Moreover, location seems to play an important role in Hg uptake by plants. In this context, Egler et al. [54] investigated Hg uptake by plants growing in the vicinity of two small-scale gold mines, where at one site, the plant Hg content was correlated with the Hg content in the soil, and at the other site, these relationships were not confirmed, indicating atmospheric deposition as an important source of aboveground biomass contamination.

Higueras et al. [60] observed differences in Hg uptake by *Olea europea* leaves at a former cinnabar mine site compared to those sampled in the vicinity of a chlorine alkali plant still in operation. They found significant relationships between soil and leaf Hg contents at the historical mining area and no soil–leaf relationships in the active chlorine alkali plant area. They explained this different uptake pattern by the predominant atmospheric Hg uptake close to the active facility in contrast to the root uptake of Hg at the historical mining area. Therefore, the age and state of the mining area play an important role in potential Hg bioaccessibility, and the data recorded in this study fully agree with these findings.

Although plant accumulation of Hg in extremely polluted areas is low compared to the Hg content in soil, the Hg content in plants exceeds that of those growing in unpolluted soil. Polluted plant biomass can pose a potential risk for herbivores living in the area. Limit values of Hg content for wildlife plants are not set in the Czech Republic, but the threshold limits for livestock feedstuffs can be used in this case. Directive No. 2002/32/ES (ES 2002) [61] defined a maximum Hg value in raw feedstuffs of 0.1 mg/kg at a biomass moisture of 12%. Therefore, recalculating the results obtained in 100% dry matter to 12% moisture, 31% of aboveground biomass samples and 83% of root samples exceeded this limit at Jedová Hora. At Horní Luby, 50% of the aboveground biomass samples and 97% of the root samples exceeded this limit. Therefore, the potential expansion of Hg pollution via the terrestrial food chain cannot be excluded.

## 4. Conclusions

The results showed high variability of the soil Hg content at the investigated locations, where both total Hg and $GEM_{soil\text{-}air}$ values varied widely, with the possible occurrence of hotspots. The sources of Hg in the hotspots should be investigated in further research, because the plant Hg contents reflected the Hg contents in the soil. Apart from $GEM_{soil\text{-}air}$, the results showed that Hg can enter the surrounding environment via plant accumulation. The spontaneously occurring plant communities growing at the locations are predominantly excluders with limited Hg accumulation ability, but the high soil Hg content resulted in elevated plant Hg levels, which differed according to the individual families and species. Among these families, the exceptional position of Fabaceae should be highlighted for their enhanced Hg accumulation ability.

In this study, the plant Hg content correlated significantly with the total Hg content in soil at the Jedová Hora location for both roots and aboveground biomass, regardless of the detectable $GEM_{soil\text{-}air}$ concentration, indicating the predominant soil–root pathway of Hg uptake, as already mentioned. At the Horní Luby location, neither roots nor aboveground biomass were significantly ($p < 0.05$) correlated with soil Hg, indicating the presence of additional factors affecting Hg uptake by plants, such as the different compositions of the plant community at individual locations, the different responses of individual plant species to soil Hg, variability in the soil Hg content within individual locations, and presumed differences in the soil characteristics between individual locations. The results showed that former mining areas deserve attention, even though they have been abandoned for a long time, because the potential risk of pollution in the surrounding environment cannot be underestimated.

**Supplementary Materials:** The following supporting information can be downloaded at https://www.mdpi.com/article/10.3390/min14121211/s1, Table S1: Description of the sampling points and total Hg contents; Table S2: Average mercury contents in the shoots and roots of the individual plant species at the Jedová Hora location; Table S3: Average mercury contents in the shoots and roots of the individual plant species at the Horní Luby location; Table S4: Spearman's ρ correlation coefficients of the interrelationships of the individual variables.

**Author Contributions:** Conceptualization, J.S. (Jiřina Száková); methodology, L.S.; investigation, M.B., J.N. and J.S. (Jiřina Sysalová); writing—original draft, M.B. and J.S. (Jiřina Száková); writing—review and editing, J.S. (Jiřina Száková) and J.S. (Jiřina Sysalová); supervision, J.S. (Jiřina Száková) and P.T. All authors have read and agreed to the published version of the manuscript.

**Funding:** The authors thank the European Regional Development Fund—Project No. CZ.02.1.01/0.0/0.0/16_019/0000845 for the financial support.

**Data Availability Statement:** The original contributions presented in this study are included in the article and Supplementary Materials.

**Conflicts of Interest:** The authors declare no conflicts of interest.

## References

1. Gworek, B.; Dmuchowski, W.; Baczewska-Dąbrowska, A.H. Mercury in the terrestrial environment: A review. *Environ. Sci. Eur.* **2020**, *32*, 128. [CrossRef]
2. Maqsood, M.; Zahra, N.; Kausar, A.; Shahzad, S.; Batool, A.; Naseer, R. Mercury Contamination and It's Dynamics in Soil–Plant Systems. In *Mercury Toxicity Mitigation: Sustainable Nexus Approach. Earth and Environmental Sciences Library*; Kumar, N., Ed.; Springer: Cham, Switzerland, 2024; pp. 45–63.
3. Selin, N.E. Global biogeochemical cycling of mercury: A review. *Annu. Rev. Environ. Resour.* **2009**, *34*, 43–63. [CrossRef]
4. Bloom, S.; Porcella, D.B. Less mercury? *Nature* **1994**, *367*, 694. [CrossRef]
5. Fernandez-Martinez, R.; Loredo, J.; Ordonez, A.; Rucandio, I. Mercury availability by operationally defined fractionation in granulometric distributions of soils and mine wastes from an abandoned cinnabar mine. *Environ. Sci. Process. Impacts* **2014**, *16*, 1069–1075. [CrossRef] [PubMed]
6. Liu, Y.R.; Zheng, Y.M.; Shen, J.P.; Zhang, L.M.; He, J.Z. Effects of mercury on the activity and community composition of soil ammonia oxidizers. *Environ. Sci. Pollut. Res.* **2010**, *17*, 1237–1244. [CrossRef]
7. Mathema, V.B.; Thakuri, B.C.; Sillanpää, M. Bacterial mer operon-mediated detoxification of mercurial compounds: A short review. *Arch. Microbiol.* **2011**, *193*, 837–844. [CrossRef]
8. Heeraman, D.A.; Claassen, V.P.; Zasoski, R.J. Interaction of lime, organic matter and fertilizer on growth and uptake of arsenic and mercury by Zorro fescue (*Vulpia myuros* L.). *Plant Soil* **2001**, *234*, 215–231. [CrossRef]
9. Moreno, F.N.; Anderson, C.W.N.; Stewart, R.B.; Robinson, B.H.; Nomura, R.; Ghomshei, M.; Meech, J.A. Effect of thioligands on plant-Hg accumulation and volatilisation from mercury-contaminated mine tailings. *Plant Soil* **2005**, *275*, 233–246. [CrossRef]

10. Linde, M.; Öborn, I.; Gustafsson, J.P. Effects of changed soil conditions on the mobility of trace metals in moderately contaminated urban soils. *Water Air Soil Pollut.* **2007**, *183*, 69–83. [CrossRef]
11. do Nascimento, F.H.; Masini, J.C. Influence of humic acid on adsorption of Hg(II) by vermiculite. *J. Environ. Manag.* **2014**, *143*, 1–7. [CrossRef] [PubMed]
12. Pei, P.; Sun, T.; Xu, Y.; Sun, Y. Soil aggregate–associated mercury (Hg) and organic carbon distribution and microbial community characteristics under typical farmland–use types. *Chemosphere* **2021**, *275*, 129987. [CrossRef] [PubMed]
13. Bloom, N.S.; Gill, G.A.; Cappellino, S.; Dobbs, C.; McShea, L.; Driscoll, C.; Mason, R.; Rudd, J. Sediment—water fluxes of mercury in Lavaca Bay, Texas. *Environ. Sci. Technol.* **1999**, *33*, 7–13. [CrossRef]
14. Reis, A.T.; Rodrigues, S.M.; Davidson, C.M.; Pereira, E.; Duarte, A.C. Extractability and mobility of mercury from agricultural soils surrounding industrial and mining contaminated areas. *Chemosphere* **2010**, *81*, 1369–1377. [CrossRef] [PubMed]
15. Beckers, F.; Rinklebe, J. Cycling of mercury in the environment: Sources, fate, and human health implications: A review. *Crit. Rev. Environ. Sci. Technol.* **2017**, *47*, 693–794. [CrossRef]
16. Shahid, M.; Khalid, S.; Bibi, I.; Bundschuh, J.; Niazi, N.K.; Dumat, C. A critical review of mercury speciation, bioavailability, toxicity and detoxification in soil-plant environment: Ecotoxicology and health risk assessment. *Sci. Total Environ.* **2020**, *711*, 134749.
17. Molina, J.A.; Oyarzun, R.; Esbrí, J.M.; Higueras, P. Mercury accumulation in soils and plants in the Almadén mining district, Spain: One of the most contaminated sites on Earth. *Environ. Geochem. Health* **2006**, *28*, 487–498. [CrossRef]
18. Ruggiero, P.; Terzano, R.; Spagnuolo, M.; Cavalca, L.; Colombo, M.; Andreoni, V.; Rao, M.A.; Perucci, P.; Monaci, E. Hg bioavailability and impact on bacterial communities in a long-term polluted soil. *J. Environ. Monit.* **2011**, *13*, 145–156. [CrossRef] [PubMed]
19. Moreno-Jiménez, E.; Peñalosa, J.M.; Esteban, E.; Carpena-Ruiz, R.O. Mercury accumulation and resistance to mercury stress in *Rumex induratus* and *Marrubium vulgare* grown in perlite. *J. Plant Nutr. Soil Sci.* **2007**, *170*, 485–494. [CrossRef]
20. Moreno-Jiménez, E.; Gimeno, H.; Gamarra, R.; Esteban, E. Evidence of a new Hg-tolerant ecotype of Rumex induratus from Almadén (Ciudad Real, Spain). *Plant Biosyst.* **2014**, *148*, 58–63. [CrossRef]
21. Xu, J.; Garcia Bravo, A.; Lagerkvist, A.; Bertilsson, S.; Sjöblom, R.; Kumpiene, J. Sources and remediation techniques for mercury contaminated soil. *Environ. Int.* **2015**, *74*, 42–53. [CrossRef]
22. Duan, P.; Khan, S.; Ali, N.; Shereen, M.A.; Siddique, R.; Ali, B.; Iqbal, H.M.N.; Nabi, G.; Sijjad, W.; Bilal, M. Biotransformation fate and sustainable mitigation of a potentially toxic element of mercury from environmental matrices. *Arab. J. Chem.* **2020**, *13*, 6949–6965. [CrossRef]
23. Li, W.C.; Tse, H.F. Health risk and significance of mercury in the environment. *Environ. Sci. Pollut. Res.* **2015**, *22*, 192–201. [CrossRef] [PubMed]
24. Raj, D.; Maiti, S.K. Sources, toxicity, and remediation of mercury: An essence review. *Environ. Monit. Assess.* **2019**, *191*, 566. [CrossRef] [PubMed]
25. Alcantara, H.J.P.; Doronila, A.I.; Kolev, S.D. Phytoextraction potential of *Manihot esculenta* Crantz. (cassava) grown in mercury- and gold-containing biosolids and mine tailings. *Miner. Eng.* **2017**, *114*, 57–63. [CrossRef]
26. García-Mercado, H.D.; Fernández-Villagómez, G.; Garzón-Zúñiga, M.A.; Durán-Domínguez-de-Bazúa, M.C. Fate of mercury in a terrestrial biological lab process using *Polypogon monspeliensis* and *Cyperus odoratus*. *Int. J. Phytoremed.* **2019**, *21*, 1170–1178. [CrossRef] [PubMed]
27. Wiafe, S.; Buamah, R.; Jackson, A.N. Phytoremediation of heavy metals contaminated water and soils from artisanal mining enclave using *Heliconia psittacorum*. *Model Earth Syst. Environ.* **2022**, *8*, 591–600.
28. Velebil, D. Cinnabar mining in Horní Luby near Cheb, Czech Republic. *Bull. Mineral.-Petrol. Oddel. Nar. Muz. Praze* **2009**, *17*, 39–61. (In Czech)
29. Velebil, D. Jedová hora Hill (Dědova hora Hill, Giftberg) near Neřežín, Czech Republic. *Bull. Mineral.-Petrol. Oddělení Národního Muz. Praze* **2003**, *11*, 86–99. (In Czech)
30. Velebil, D.; Losos, Z. Mercury rich tetraedrite from Jedová hora Hill near Neřežín and its accompanying minerals. *Bull. Mineral.-Petrol. Odděleni Nar. Muz. Praze* **2008**, *16*, 56–60. (In Czech)
31. Hojdová, M.; Navrátil, T.; Rohovec, J.; Penížek, V.; Grygar, T. Mercury distribution and speciation in soils affected by historic mercury mining. *Water Air Soil Pollut.* **2009**, *200*, 89–99. [CrossRef]
32. Sysalová, J.; Kučera, J.; Drtinová, B.; Červenka, R.; Zvěřina, O.; Komárek, J.; Kameník, J. Mercury species in formerly contaminated soils and released soil gases. *Sci. Total Environ.* **2017**, *584–585*, 1032–1039. [CrossRef] [PubMed]
33. Száková, J.; Burešová, A.; Praus, L.; García-Sánchez, M.; Holečková, Z.; Gabriel, J.; Sysalová, J.; Červenka, R.; Komárek, J.; Grohová, S.; et al. The response of mercury (Hg) transformation in soil to sulfur compounds and sulfur-rich biowaste application. *Environ. Earth Sci.* **2016**, *75*, 584. [CrossRef]
34. Száková, J.; Kolihová, D.; Miholová, D.; Mader, P. Single-purpose atomic absorption spectrometer AMA-254 for mercury determination and its performance in analysis of agricultural and environmental materials. *Chem. Pap.* **2004**, *58*, 311–315.
35. Hu, B.F.; Jia, X.L.; Hu, J.; Xu, D.Y.; Xia, F.; Li, Y. Assessment of heavy metal pollution and health risks in the soil-plant-human system in the Yangtze river delta, China. *Int. J. Environ. Res. Public Health* **2017**, *14*, 1042. [CrossRef]

36. *Public Notice No.153/2016 (2016) About the Conditions for the Protection of the Agricultural Soil Quality*; Legal Code of the Czech Republic: Brno, Czech Republic, 2016; pp. 2692–2699.
37. Cibulka, J.; Domažlická, E.; Kozák, J. *Transfer of Lead, Cadmium and Mercury in the Biosphere*; Academia Praha: Prague, Czech Republic, 1991; p. 432. (In Czech)
38. Fernández-Martínez, R.; Larios, R.; Gómez-Pinilla, I.; Gómez-Mancebo, B.; López-Andrés, S.; Loredo, J.; Ordóñez, A.; Rucandio, I. Mercury accumulation and speciation in plants and soils from abandoned cinnabar mines. *Geoderma* **2015**, *253–254*, 30–38. [CrossRef]
39. de la Iglesia-Turiño, S.; Febrero, A.; Jauregui, O.; Caldelas, C.; Araus, J.L.; Bort, J. Detection and quantification of unbound Phytochelatin 2 (PC2) in plant extracts of Brassica napus grown with different levels of mercury. *Plant Physiol.* **2006**, *142*, 742–749. [CrossRef]
40. Fernández, S.; Poschenrieder, C.; Marcenò, C.; Gallego, J.R.; Jiménez-Gámez, D.; Bueno, A.; Afif, E. Phytoremediation capability of native plant species living on Pb-Zn and Hg-As mining wastes in the Cantabrian range, north of Spain. *J. Geochem. Exp.* **2017**, *174*, 10–20. [CrossRef]
41. Raj, D. Bioaccumulation of mercury, arsenic, cadmium, and lead in plants grown on coal mine soil. *Hum. Ecol. Risk Assess. Int. J.* **2019**, *25*, 3. [CrossRef]
42. Millán, R.; Lominchar, M.A.; Rodríguez-Alonso, J.; Schmid, T.; Sierra, M.J. Riparian vegetation role in mercury uptake (Valdeazogues River, Almadén, Spain). *J. Geochem. Exp.* **2014**, *140*, 104–110. [CrossRef]
43. Esteban, E.; Moreno, E.; Peñalosa, J.; Cabrero, J.I.; Millán, R.; Zornoza, P. Short and long-term uptake of Hg in white lupin plants: Kinetics and stress indicators. *Environ. Exp. Bot.* **2008**, *62*, 316–322. [CrossRef]
44. Carrasco-Gil, S.; Estebaranz-Yubero, M.; Medel-Cuesta, D.; Millán, R.; Hernández, L.E. Influence of nitrate fertilization on Hg uptake and oxidative stress parameters in alfalfa plants cultivated in a Hg-polluted soil. *Environ. Exp. Bot.* **2012**, *75*, 16–24. [CrossRef]
45. Nonnoi, F.; Chinnaswamy, A.; de la Torre, V.S.G.; de la Peña, T.C.; Lucas, M.M.; Pueyo, J.J. Metal tolerance of rhizobial strains isolated from nodules of herbaceous legumes (*Medicago* spp. and *Trifolium* spp.) growing in mercury-contaminated soils. *Appl. Soil Ecol.* **2012**, *61*, 49–59. [CrossRef]
46. Raghupathy, S.; Arunachalam, S. Trends in legume-rhizobia symbiosis in remediation of mercury-contaminated agricultural soils. *Commun. Soil Sci. Plant Anal.* **2024**, *55*, 916–930. [CrossRef]
47. Dary, M.; Chamber-Pérez, M.A.; Palomares, A.J.; Pajuelo, E. "In situ" phytostabilisation of heavy metal polluted soils using Lupinus luteus inoculated with metal resistant plantgrowth promoting rhizobacteria. *J. Hazard. Mater.* **2010**, *177*, 323–330. [CrossRef]
48. de la Torre, V.S.G.; de la Peña, T.C.; Lucas, M.M.; Pueyo, J.J. Rapid screening of Medicago truncatula germplasm for mercury tolerance at the seedling stage. *Environ. Exp. Bot.* **2013**, *91*, 90–96. [CrossRef]
49. Su, Y.; Han, F.X.; Chen, J.; Sridhar, B.B.M.; Monts, D.L. Phytoextraction and accumulation of mercury in three plant species:: Indian mustard (*Brassica juncea*), beard grass (*Polypogon monospeliensis*), and Chinese brake fern (*Pteris vittata*). *Int. J. Phytoremed.* **2008**, *10*, 547–560. [CrossRef]
50. Wang, Z.; Wang, H.; Wang, H.; Li, Q.; Li, Y. Heavy metal pollution and potential health risks of commercially available Chinese herbal medicines. *Sci. Total Environ.* **2019**, *653*, 748–757. [CrossRef]
51. Yang, B.; Gao, Y.; Zhang, C.; Zheng, X.; Li, B. Mercury accumulation and transformation of main leaf vegetable crops in Cambosol and Ferrosol soil in China. *Environ. Sci. Pollut. Res.* **2020**, *27*, 391–398. [CrossRef]
52. Peralta-Videa, J.R.; Lopez, M.L.; Narayan, M.; Saupe, G.; Gardea-Torresdey, J. The biochemistry of environmental heavy metal uptake by plants: Implications for the food chain. *Int. J. Biochem. Cell Biol.* **2009**, *41*, 1665–1677. [CrossRef]
53. Umlaufová, M.; Száková, J.; Najmanová, J.; Sysalová, J.; Tlustoš, P. The soil-plant transfer of risk elements within the area of an abandoned gold mine in Libčice, Czech Republic. *J. Environ. Sci. Health A Tox. Hazard. Subst. Environ. Eng.* **2018**, *53*, 1267–1276. [CrossRef]
54. Egler, S.G.; Rodrigues-Filho, S.; Villas-Bôas, R.C.; Beinhoff, C. Evaluation of mercury pollution in cultivated and wild plants from two small communities of the Tapajós gold mining reserve, Pará State, Brazil. *Sci. Total Environ.* **2006**, *368*, 424–433. [CrossRef] [PubMed]
55. Sasmaz, M.; Akgül, B.; Yıldırım, D.; Sasmaz, A. Mercury uptake and phytotoxicity in terrestrial plants grown naturally in the Gumuskoy (Kutahya) mining area, Turkey. *Int. J. Phytoremediat.* **2016**, *18*, 69–76. [CrossRef] [PubMed]
56. Tipping, E.; Wadsworth, R.A.; Norris, D.A.; Hall, J.R.; Ilyin, I. Long-term mercury dynamics in UK soils. *Environ. Pollut.* **2011**, *159*, 3474–3483. [CrossRef] [PubMed]
57. Meloni, F.; Farieri, A.; Higueras, P.L.; Esbrí, J.M.; Nisi, B.; Cabassi, J.; Rappuoli, D.; Vaselli, O. Mercury distribution in plants and soils from the former mining area of Abbadia San Salvatore (Tuscany, Central Italy). *Environ. Geochem. Health* **2023**, *45*, 8523–8538. [CrossRef]
58. García-Sánchez, A.; Murciego, A.; Álvarez-Ayuso, E.; Santa Regina, I.; Rodríguez-González, M.A. Mercury in soils and plants in an abandoned cinnabar mining area (SW Spain). *J. Hazard. Mater.* **2009**, *168*, 1319–1324. [CrossRef]
59. Lindberg, S.E.; Jackson, D.R.; Huckabee, J.W.; Janzen, S.A.; Levin, M.J.; Lund, J.R. Atmospheric emission and plant uptake of mercury from agricultural soils near the Almaden mercury mine. *J. Environ. Qual.* **1979**, *8*, 572–578. [CrossRef]

60. Higueras, P.L.; Amorós, J.Á.; Esbrí, J.M.; Pérez-de-los-Reyes, C.; López-Berdonces, M.A.; García-Navarro, F.J. Mercury transfer from soil to olive trees. A comparison of three different contaminated sites. *Environ. Sci. Pollut. Res.* **2016**, *23*, 6055–6061. [CrossRef]
61. *Directive No. 2002/32/ES of European Parliament and Council of Europe Concerning Xenobiotics in Feedstuffs*; European Parliament and Council of Europe: Strasbourgh, France, 2002.

minerals

MDPI

Article

# Comparative Analysis of Reduction Techniques Aiming for the Minimization of Contaminated Soil with Red Mud

Srećko Stopić [1,*], Duško Kostić [1,2], Mitar Perušić [2], Richard Schneider [1], Isnaldi R. Souza Filho [3,4], Aleksandar Mitrašinović [5] and Bernd Friedrich [1]

1 IME Process Metallurgy and Metal Recycling, RWTH Aachen University, 52056 Aachen, Germany; dkostic@ime-aachen.de (D.K.); bfriedrich@ime-aachen.de (B.F.)
2 Faculty of Technology Zvornik, University of East Sarajevo, 75400 Zvornik, Republic of Srpska, Bosnia and Herzegovina; mitar.perusic@tfzv.ues.rs.ba
3 Max-Planck-Institut for Sustainable Materials, Max-Planck-Strasse 1, 40237 Duesseldorf, Germany
4 Institut Jean Lamour, CNRS (UMR 7198), Université de Lorraine, F-54000 Nancy, France
5 Institute of Technical Sciences of the Serbian Academy of Sciences and Arts, 11000 Belgrade, Serbia; aleksandar.mitrasinovic@itn.sanu.ac.rs
* Correspondence: sstopic@ime-aachen.de

**Abstract:** Comparative analysis of red mud reduction techniques was performed using both carbothermal and hydrogen-based reduction methods, combining thermochemical modeling and experimental validation. The reduction process is mostly important because of the high contamination risk assessment of soil with disposed red mud. Therefore, the minimization of red mud during the reduction process can be a novel strategy for the production of metallic iron and solid residue for hydrometallurgical treatment. Different strategies of hydrogen and carbon reduction in static and dynamic conditions were studied between 700 °C and 1700 °C. The separation of solid residue and formed iron was analyzed using magnetic separation. The main aim was to study the advantages and disadvantages of using decarbonizing technologies for the treatment of red mud, aiming to develop an environmentally friendly process. Thermochemical analysis of the reduction offered new data about mass losses during our process through the evaporation, thermal decomposition, and formation of metallic carbide.

**Keywords:** reduction; red mud; hydrogen; carbon

## 1. Introduction

The reduction of iron oxides is a fundamental process in metallurgy, particularly in the production of iron and steel. It involves the removal of oxygen from iron oxides, such as hematite ($Fe_2O_3$), magnetite ($Fe_3O_4$), and wüstite (FeO), to obtain metallic iron. This transformation is typically achieved through chemical reactions with reducing agents like carbon, carbon monoxide, or hydrogen [1].

In industrial settings, the most common method is the blast furnace process, where iron ore is reduced using coke at high temperatures, producing molten iron that is later refined into steel. An alternative approach, known as direct reduction, employs natural gas or hydrogen to convert iron oxides into solid sponge iron at lower temperatures. Hydrogen plasma reduction is another promising method that uses ionized hydrogen gas to achieve reduction at high efficiency and with minimal carbon emissions. Additionally, red mud, a byproduct of aluminum production rich in iron oxides, can be processed through reduction techniques to extract iron, making use of industrial waste. There is also

 https://doi.org/10.3390/min15050470

the thermite process, which uses aluminum to reduce iron oxide in a highly exothermic reaction [1–4].

- Carbothermic reduction

Carbothermic reduction is the primary method for extracting metals from ores using carbon as a reducing agent, playing a crucial role in iron and steel production. This process involves the reaction of iron oxides, such as hematite ($Fe_2O_3$) and magnetite ($Fe_3O_4$), with carbon (coke or charcoal) at high temperatures, typically in a blast furnace around 1500 °C. As carbon removes oxygen from the ore, it forms carbon monoxide (CO), which also reacts with iron oxides, producing carbon dioxide ($CO_2$) and metallic iron [5,6].

The choice of reduction method depends on factors such as energy efficiency, cost, and environmental impact. With increasing concerns about carbon emissions, newer techniques focusing on hydrogen-based reduction, hydrogen plasma, and carbon-neutral approaches are being explored to make the process more sustainable for the future of iron and steel production.

$$3Fe_2O_3 + CO \rightarrow 2Fe_3O_4 + CO_2 \quad (1)$$

Then, magnetite is reduced to wüstite (FeO) [6] as follows:

$$Fe_3O_4 + CO \rightarrow 3FeO + CO_2 \quad (2)$$

Finally, wüstite (FeO) is reduced to metallic iron [6] as follows:

$$FeO + CO \rightarrow Fe + CO_2 \quad (3)$$

Carbothermic reduction is an endothermic process requiring energy, which is supplied by carbon combustion in the presence of oxygen. This generates carbon monoxide, which, along with coke, acts as a reducing agent, forming the basis of pig iron production. Pig iron contains carbon and impurities, necessitating further refining for pure steel. A similar process applies to red mud, as it contains iron (III) oxide, like iron ore. Efficiency depends on factors such as carbon reactivity, furnace temperature and pressure, and ore characteristics. Optimal conditions ensure complete reduction while minimizing slag formation. Thermodynamically, Gibbs free energy favors a reduction of metallic oxides at higher temperatures, while reaction kinetics influence overall efficiency [7–9].

- Hydrogen reduction

Carbon remains the dominant reducing agent in global metal production, with carbothermic reduction considered the most cost-effective method. However, the industry is under pressure to reduce carbon emissions, and hydrogen is emerging as a promising alternative. The European Union aims to cut industrial carbon emissions by up to 95% by 2050, while Australia is actively promoting hydrogen as a long-term solution, with initiatives like the Hydrogen Working Group supporting clean hydrogen adoption by 2030 [10].

Hydrogen offers significant environmental benefits, reducing $CO_2$ emissions by up to 95%, with water as its only byproduct. However, its high production costs, energy-intensive generation (especially for green hydrogen), and safety concerns pose major challenges. Large-scale adoption requires significant expansion in production capacity, as well as advancements in storage and transport. Currently, less than 10% of global hydrogen is used in metal production, mainly for refractory metal powders like tungsten, molybdenum, and some iron and nickel [11].

The reduction of hematite by hydrogen proceeds in two or three steps, via magnetite ($Fe_3O_4$) and wüstite (FeO), according to the following equations [3]:

$$3Fe_2O_3 + H_2 \rightarrow 2Fe_3O_4 + H_2O \quad (4)$$

$$Fe_3O_4 + H_2 \rightarrow 3FeO + H_2O \quad (5)$$

$$FeO + H_2 \rightarrow Fe + H_2O \quad (6)$$

Despite its potential, hydrogen reduction faces industrial limitations. Its weaker reducing ability compared to carbon at high temperatures, the endothermic nature of the process, and the need for reactor redesigns complicate large-scale implementation. Additionally, hydrogen interactions with impurities can affect efficiency, while high operational costs remain a barrier to widespread adoption [12].

- Hydrogen plasma reduction

In addition to direct hydrogen reduction, an innovative approach involves using plasma hydrogen to reduce metallic oxides. Plasma hydrogen provides kinetic and thermodynamic advantages due to its highly reactive ionic and vibrationally excited species, which transfer energy efficiently to alter raw material structures. This localized heating reduces the need for bulk heating, improving overall energy efficiency [13].

Plasma, the fourth state of matter, forms when high energy ionizes a gas. Industrial plasmas are classified as thermal and non-thermal, with thermal plasmas generated through electrode discharge, radiofrequency, or microwave fields at very low pressures. High electron temperatures and rapid collisions among species enable quick local thermodynamic equilibrium, making plasma hydrogen a promising alternative for metal reduction [13].

Hydrogen plasma reduction is an advanced metallurgical technique for reducing iron oxides, mainly hematite, into metallic iron. Plasma is created by supplying high energy to hydrogen gas, often mixed with argon, ionizing hydrogen molecules through electrical discharge, radiofrequency, or microwave fields. This generates a reactive plasma containing ions ($H^+$), electrons, atomic hydrogen (H), and excited molecular species, like $H_2^*$ and $H_3^+$ [14].

When hematite ($Fe_2O_3$) interacts with hydrogen plasma, its high-energy components initiate reduction. Highly reactive atomic hydrogen and hydrogen ions first convert hematite to magnetite ($Fe_3O_4$), marking the initial step in the reduction process [15] as follows:

$$3Fe_2O_3 + 2H \rightarrow 2Fe_3O_4 + H_2O \quad (7)$$

and then a reduction of magnetite to wüstite

$$Fe_3O_4 + 2H \rightarrow 3FeO + H_2O \quad (8)$$

and finally, to metallic iron (Fe)

$$FeO + 2H \rightarrow Fe + H_2O \quad (9)$$

Hydrogen plasma reduction offers advantages over conventional hydrogen and carbon-based methods, including localized heating for better energy efficiency and faster reduction due to highly reactive plasma species. It also produces only water vapor, making it environmentally friendly compared to carbothermic reduction [14].

However, challenges include maintaining stable plasma conditions, controlling energy input, and preventing unwanted reactions. Optimizing temperature, pressure, and plasma composition is essential for efficiency. Despite these hurdles, hydrogen plasma reduction holds great potential for sustainable iron production, with future research focusing on scaling up and improving plasma generation for industrial applications [16].

- Red Mud

Red mud (RM), a byproduct of alumina production, is named for its red color due to its high iron oxide content. Typically, 0.5 to 2 tons of red mud are produced per ton of alumina, with global accumulation surpassing 5 billion tons by 2020, led by China. Red mud's complex composition includes CaO, $SiO_2$, $Fe_2O_3$, $TiO_2$, $Na_2O$, $K_2O$, and various minerals. Its high alkalinity (pH 9.0 to 13.2) is mainly due to soluble and insoluble alkalis, which can pose environmental threats like groundwater contamination, corrosion, and toxicity to organisms [17–19].

Safe disposal of red mud is a major challenge, with methods like slurry disposal, dry mud disposal, and solar drying being used. Red mud dams, designed to store large quantities of this material, help prevent leakage and control environmental risks. However, red mud's high alkalinity, potential heavy metal contamination, and fine particles pose significant environmental and health risks, such as soil and water pollution, corrosion, and air pollution. The Kolontar disaster in Hungary highlighted the catastrophic potential of improper red mud management, causing environmental damage, injuries, and loss of life. Proper management is crucial to mitigate these risks [20–23].

Thermochemical modeling has been conducted to analyze the reduction of red mud using both carbon and hydrogen as reducing agents. Red mud, a byproduct of the aluminum industry, contains a significant amount of iron oxides, making it a potential secondary source for iron extraction. The reduction process has been studied using FactSage 8.0, a powerful thermochemical software, to simulate reaction mechanisms, equilibrium compositions, and process efficiency under different conditions.

For carbon-based reduction, modeling evaluates how coke or charcoal interacts with iron oxides in red mud, examining temperature-dependent reaction pathways and the formation of byproducts, such as slag and $CO_2$. Similarly, hydrogen reduction modeling focuses on the efficiency of hydrogen gas in converting iron oxides to metallic iron, analyzing reaction kinetics, temperature influence, and the environmental benefits of water vapor as the only byproduct.

The novelty of this paper lies in the comprehensive analysis of various reduction methods for extracting iron from red mud, with a focus on carbon and hydrogen reduction in both tubular furnaces and rotary kilns, as well as hydrogen plasma reduction. This study not only explores conventional reduction methods but also introduces the potential of hydrogen plasma reduction, which is a more sustainable and energy-efficient approach. Furthermore, the paper employs advanced thermochemical modeling using FactSage to simulate and compare the reaction mechanisms, equilibrium compositions, and process efficiencies of these reduction methods under various conditions. By combining experimental approaches with detailed modeling, this paper provides valuable insights into optimizing reduction processes for red mud, offering a thorough comparison of traditional and innovative techniques for metal extraction, with particular attention to environmental impact and scalability.

## 2. Materials and Methods

Many advanced analytical techniques were employed to accurately characterize and analyze the materials involved in this research. Each of these methods provides distinct insights into the structure, composition, and size of the samples. Used techniques are X-ray diffraction (XRD-Bruker D8 Advance with LynxEye detector, Billerica, MA, USA), Energy-Dispersive X-ray Spectroscopy (EDS-Octane Plus-A detector by Ametek-EDAX, Berwyn, PA, USA), Scanning Electron Microscopy (SEM-JSM 7000F by JEOL (2006 model,

JEOL Ltd., Tokyo, Japan)), Inductively Coupled Plasma Optical Emission Spectroscopy (ICP-OES Agilent 5000, Santa Clara, CA, USA), RWTH, and Aachen.

In this work, red mud from the factory "Alumina" Ltd. Zvornik (Zvornik, Republic of Srpska, Bosnia and Herzegovina) was used as a raw material. The red mud is previously dried, grounded, and prepared for the reduction process. In Table 1, the chemical composition of the used red mud is shown.

**Table 1.** Composition of red mud from "Alumina" Ltd. Zvornik.

| Compounds | Wt., % | Compounds | Wt., % |
|---|---|---|---|
| Ignition loss at 1000 °C | 8.32 | $Ga_2O_3$ | 0.225 |
| $SiO_2$ | 10.52 | CuO | 0.007 |
| $Fe_2O_3$ | 49.29 | $K_2O$ | 0.159 |
| $Na_2O$ | 2.45 | $Tl_2O_3$ | 0.088 |
| $TiO_2$ | 4.59 | MnO | 0.145 |
| CaO | 8.23 | MgO | 0.627 |
| $Al_2O_3$ | 12.03 | NiO | 0.034 |
| $Ag_2O$ | 0.001 | PbO | 0.019 |
| BaO | 0.014 | $P_2O_5$ | 0.930 |
| $Cr_2O_3$ | 0.133 | ZnO | 0.016 |
| $Sc_2O_3$ | 0.011 | $V_2O_5$ | 0.135 |
| $Co_2O_3$ | 0.012 | SrO | 0.075 |

Red mud characterization was performed using ICP and EDS analysis.

The XRD analysis in Figure 1 has confirmed the presence of hematite and other very stable oxides for reduction, including ilmenite. Analysis shows that other non-ferrous minerals, such as perovskite, calcite, diaspore, boehmite, anhydrite, etc., will undergo structural change caused by iron oxide reduction.

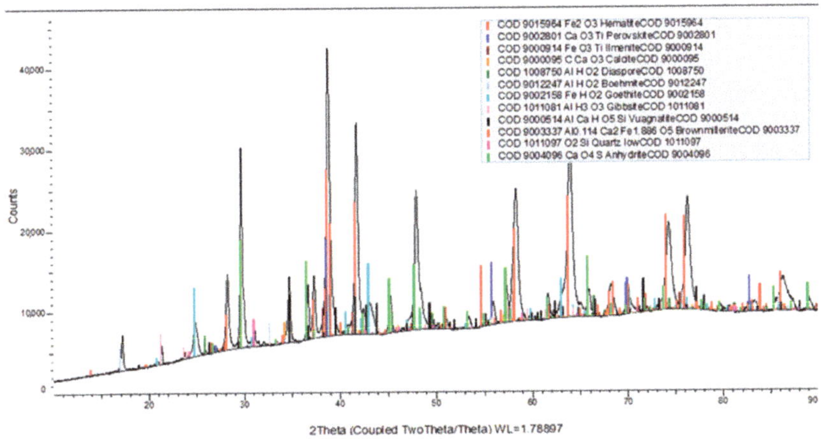


**Figure 1.** XRD analysis of red mud.

The reduction process with carbon took place in an electric arc furnace (carbothermal reduction), the reduction process with hydrogen was carried out in a tubular furnace (H-tubular reduction) and a rotary kiln (H-rotary kiln reduction), and hydrogen plasma reduction was carried out at the Max Plank Institute for Sustainable Materials in Dusseldorf (H-plasma reduction). The reduced samples were separated mechanically using magnetic separation. The obtained products were analyzed by XRD and EDS analysis as well as SEM analysis. In Table 2, the design of the reduction experiments is shown.

**Table 2.** Design of the reduction experiments.

| Exp. No | Reducing Agent | Temperature (°C) | Time (min) | Mass or Flow of Reducing Agent |
|---|---|---|---|---|
| 1 | Carbon (EAF) | 1600 | 60 | 10 g |
| 2 | Hydrogen (rotary kiln) | 920 | 120 | 2 L/min |
| 3 | Hydrogen (tubular furnace) | 800 | 60 | 3 L/min (2 $H_2$/1Ar) |
| 4 | Hydrogen (tubular furnace) | 900 | 60 | 3 L/min (2 $H_2$/1Ar) |
| 5 | Hydrogen (tubular furnace) | 1000 | 60 | 3 L/min (2 $H_2$/1Ar) |
| 6 | Hydrogen plasma | 2500 | 5 | 10 L/min (1 $H_2$/9Ar) |
| 7 | Hydrogen plasma | 2500 | 10 | 10 L/min (1 $H_2$/9Ar) |
| 8 | Hydrogen plasma | 2500 | 15 | 10 L/min (1 $H_2$/9Ar) |

The reduction process with hydrogen as a reducing agent could not take place at a higher temperature due to equipment limitations. Hydrogen reduction equipment cannot withstand temperatures higher than 1000 °C.

### 2.1. Thermochemical Modeling

For the thermochemical analysis of the reduction of red mud at temperatures of 1600 °C and 1700 °C for different concentrations of both hydrogen and carbon, the calculation is performed without reducing agents. The thermodynamic modeling of the reduction of iron oxide from red mud is carried out using FactSage™ 8.0 software [24], developed by GTT Technologies, Kohlscheid, Germany. This modeling approach allows for a detailed analysis of phase equilibria and chemical reactions under varying conditions. The chemical composition of red mud used for calculations is a $SiO_2$ content of 12.43%. Iron oxide content is 58.22%, titanium dioxide content is 5.42%, calcium oxide is 9.72%, and alumina content is 14.21%.

A detailed overview of all input parameters and simulation conditions is presented in Table 3. In each simulation, a fixed input mass of 100 g of red mud is used to ensure consistency in comparative analysis.

**Table 3.** Design of the experiments for thermochemical modeling.

| Experiment No. | Temperature (°C) | Mass of Reducing Agent (g) |
|---|---|---|
| 1 | 1600 and 1700 | 0 (Hydrogen) |
| 2 | 1600 and 1700 | 5 (Hydrogen) |
| 3 | 1600 and 1700 | 10 (Hydrogen) |
| 4 | 1600 and 1700 | 15 (Hydrogen) |
| 5 | 1600 and 1700 | 20 (Hydrogen) |
| 6 | 1600 and 1700 | 25 (Hydrogen) |
| 7 | 1600 and 1700 | 30 (Hydrogen) |
| 8 | 1600 and 1700 | 35 (Hydrogen) |
| 9 | 1600 and 1700 | 40 (Hydrogen) |

Table 3. Cont.

| Experiment No. | Temperature (°C) | Mass of Reducing Agent (g) |
| --- | --- | --- |
| 10 | 1600 and 1700 | 45 (Hydrogen) |
| 11 | 1600 and 1700 | 50 (Hydrogen) |
| 12 | 1600 and 1700 | 0 (Carbon) |
| 13 | 1600 and 1700 | 5 (Carbon) |
| 14 | 1600 and 1700 | 10 (Carbon) |
| 15 | 1600 and 1700 | 15 (Carbon) |
| 16 | 1600 and 1700 | 20 (Carbon) |

The simulations cover hydrogen and carbon at different concentrations to examine their effects on phase transformations, metal recovery rates, and byproduct formation.

### 2.2. Carbothermic Reduction in an Electric Arc Furnace

The particle sizes of red mud have values between 4.7 and 5.98 μm. The reduction process of red mud with carbon took place in a DC electric arc furnace at 1600 °C for 90 min. Three (3) kg of red mud and 10 g of graphite per 100 g of red mud as carbon-reducing agents are mixed and charged into the furnace. Because of the high viscosity of the slag after reducing the iron oxides, the slag creates a foam. To lower the viscosity, CaO has been added as a fluxing agent. We added 100 g CaO/3 kg red mud in order to facilitate the separation of slag and formed metallic iron. The positive side effect of CaO is that it is possible that FeO from $Fe_2SiO_4$ is dissolved back into the slag with a positive change in the activity of FeO. After charging, a holding time of 45 min has been started to let the C react with the slag. After the holding time, the furnace was tapped to extract the Fe metal phase, while the slag phase remained in the crucible. The samples were separated mechanically using magnetic separation [25].

### 2.3. Hydrogen Reduction in a Tubular Furnace

A reduction of 1.0 g of red mud was performed in a small tubular furnace at temperatures 800, 900, and 1000 °C for 60 min. According to previous studies of hydrogen reduction [26–30], the reduction process utilized a hydrogen flow rate of 2 L/min and an argon flow rate of 1 L/min [31].

### 2.4. Hydrogen Reduction in a Rotary Kiln

The reduction of red mud was conducted using a Carbolite rotary kiln. The objective was to investigate the reduction behavior of red mud in a controlled hydrogen atmosphere at high temperatures. A quartz tube was employed as the reaction chamber, and the process was carried out at 920 °C with hydrogen gas flows to facilitate controlled reduction. Figure 2 shows the equipment used for the experiments.

A total of 200 g of red mud was placed inside the quartz tube, which was then securely connected to the rotary kiln. Nitrogen gas was continuously supplied throughout the heating process. Once the system reached the target temperature of 920 °C, hydrogen was introduced at a flow rate of 1 L/min for 10 min, followed by another 10 min of minimal hydrogen flow.

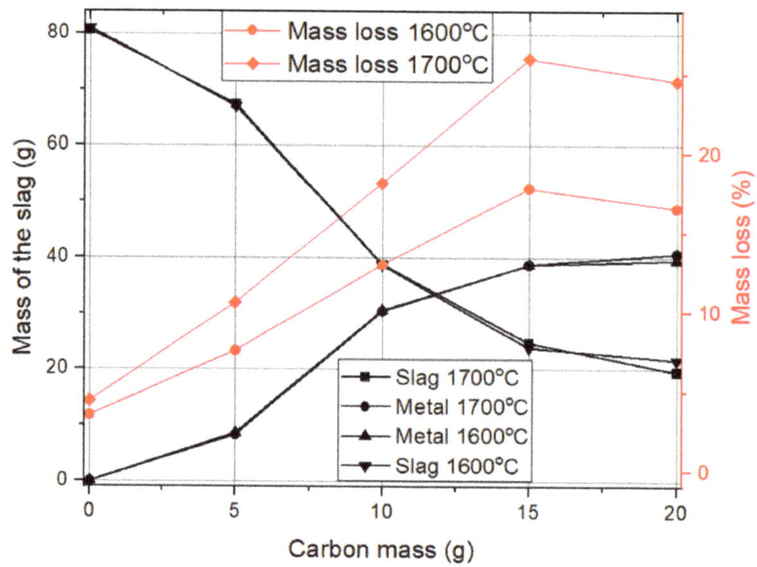


**Figure 2.** Mass loss and the mass of the slag and metal dependence on the carbon mass.

*2.5. Hydrogen Plasma Reduction in an Electric Arc Furnace*

The electric arc furnace is equipped with a tungsten electrode with a diameter of 6 mm. For the experiments, 20 g of red mud was placed into the water-cooled, copper sample holder. For the melting process, the 18-liter melting chamber is filled with a gas mixture consisting of 10% hydrogen and 90% argon. The total pressure is 900 mbar. An arc is ignited from the electrode to the red mud, which consists of plasma heated to 2350–2050 °C. The melting process is cyclical. Each cycle consists of melting the red mud with the plasma for one minute, followed by a brief cooling of the feed material. During the cooling phase, the hydrogen consumed during the reduction is replenished. The overall reduction time is evaluated.

The obtained products were analyzed by XRD, EDS analysis, and SEM analysis.

## 3. Results and Discussion

*3.1. Carbothermal Reduction Modeling*

During the reduction process, $Fe_2O_3$ is firstly reduced to FeO, and with the increase in reducing agent mass, it is further reduced to metallic Fe [32]. After the reduction process, there are two phases: slag, which is low in Fe, and the metal Fe phase. The carbothermal reduction of oxides from red mud was calculated and shown in Figure 2.

At 0 g of carbon mass, the process consists only of the decomposition of red mud at high temperatures, leading to the formation of slag without any metal phase. As the carbon mass increases, the metal content also increases, while the slag content decreases. This change is quite significant up to around 15 g of carbon mass, after which the variations are not as significant.

Regarding temperature effects, the trends in slag and metal mass are similar at both 1600 °C and 1700 °C. However, a slightly better metal yield is observed at 1700 °C, indicating a more efficient reduction process at higher temperatures. Mass loss trends show that higher temperatures result in greater mass loss, but the difference is not very significant. The overall observation suggests that increasing the carbon mass plays a crucial role in metal extraction, while temperature has a moderate but noticeable impact.

The Figure 3 illustrates the composition of alloying elements in iron during carbothermal reduction. At lower carbon mass values between 5 and 10 g, iron dominates with a content close to 100%. However, with an increase in carbon mass beyond 15 g, there is a significant rise in the content of other elements in the metal phase. Silicon content increases notably, reaching values up to 11%, while titanium and carbon contents also show substantial growth. The presence of carbon in the metal phase suggests an excess amount beyond what is needed for reduction. Temperature plays a significant role in this process, with higher carbon content leading to a greater incorporation of impurities. At 1700 °C, the content of all alloying elements in iron is higher compared to 1600 °C, indicating that increased temperature enhances impurity dissolution. This suggests that a carbon mass above 10 g leads to excessive impurity formation, which may not be desirable for obtaining high-purity iron.

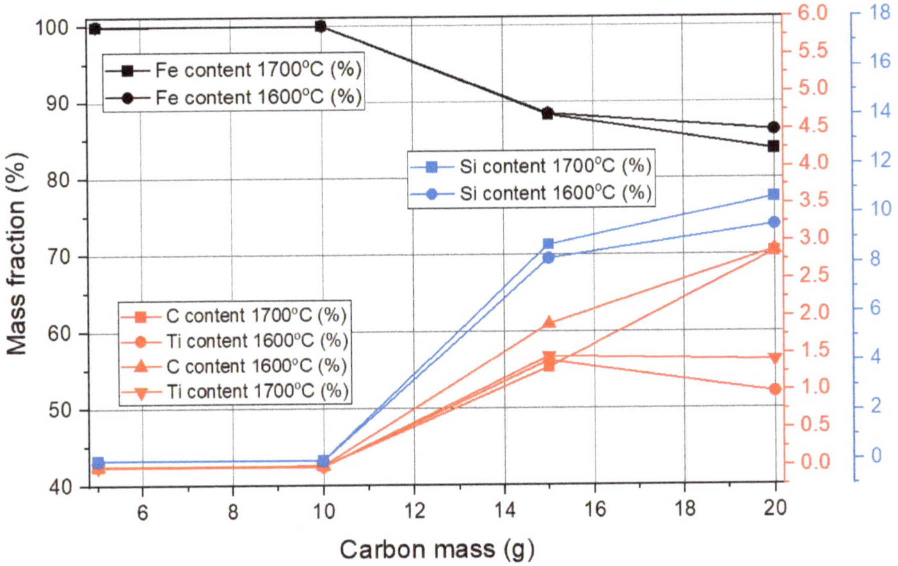


**Figure 3.** The mass fraction of elements in the metal phase correlates with the carbon mass.

Figure 4 shows the transformation of iron oxides with varying carbon masses and temperatures. The $Fe_2O_3$ content drops significantly with the addition of just 5 g of carbon, but it does not directly reduce to metallic iron. Instead, it first converts into FeO. The FeO content then decreases sharply as the carbon mass increases, reaching a very low level, close to zero, at around 15 g of carbon. This suggests that the reduction process occurs in stages, with $Fe_2O_3$ first reducing to FeO before eventually transforming into metallic iron. Additionally, at higher carbon masses, a third phase, TiC, is formed due to the excess carbon, indicating that beyond a certain point, surplus carbon leads to the formation of additional carbide phases instead of contributing to further iron reduction.

These observations suggest that the optimal carbon mass for achieving a high Fe content while minimizing excessive alloying and carbide formation lies between 10 g and 15 g. Beyond this range, further carbon additions promote carbide formation, which may influence the final properties of the metal phase.

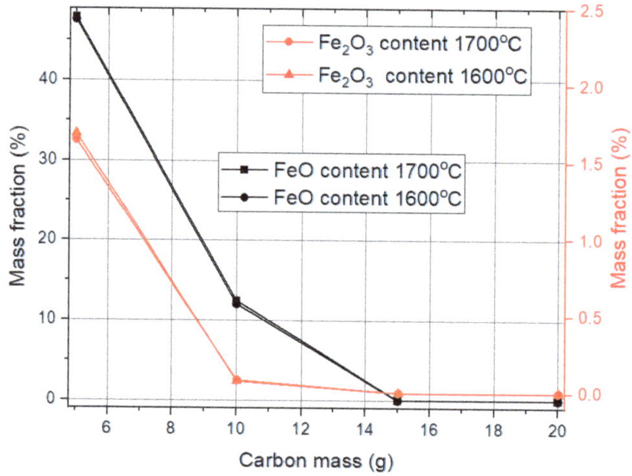


**Figure 4.** Mass fraction dependence on the carbon mass.

### 3.2. Hydrogen Reduction Modeling

The hydrogen reduction of oxides from red mud is calculated and shown in Figure 5.

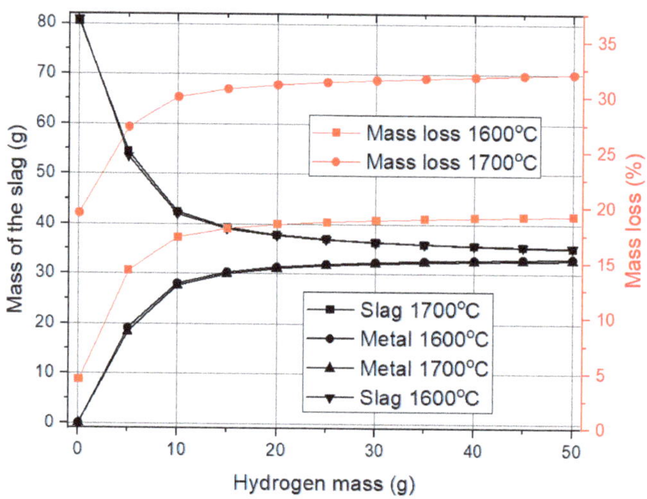


**Figure 5.** Mass loss and mass of the slag and metal dependence on the hydrogen mass.

Figure 5 shows the influence of hydrogen mass on slag mass and mass loss during the reduction process. Initially, with no hydrogen present, the system is entirely composed of red mud, resulting in the maximum slag mass. As hydrogen is introduced, reduction reactions take place, leading to a notable decrease in slag mass and a corresponding increase in metal formation. This effect is most significant for up to 20 g of hydrogen, beyond which additional hydrogen has little impact on further reducing the slag or increasing metal yield, proving that the reaction is almost complete.

Mass loss exhibits a similar pattern. With the hydrogen mass increase, oxygen from iron oxide reacts with hydrogen to form water vapor. Up to 20 g of hydrogen, mass loss increases significantly, but beyond this point, the effect becomes less significant, suggesting that most reducible oxides have already been removed. Temperature variation from 1600 °C to 1700 °C does not greatly influence slag reduction or metal formation, but it does have a noticeable effect on mass loss. At 1700 °C, mass loss is higher, likely due to enhanced reaction rates and increased volatilization of reaction byproducts.

These findings suggest that 20 g of hydrogen is an optimal threshold for the reduction process. Beyond this, additional hydrogen offers minimal benefits in terms of metal recovery and slag reduction, while promoting slightly greater mass loss at higher temperatures.

Figure 6 presents how the mass fractions of different iron-containing species change with increasing hydrogen mass. The $Fe_2O_3$ content in the slag decreases sharply from 58% to nearly zero when the hydrogen mass reaches 5 g, indicating that $Fe_2O_3$ is rapidly reduced. However, this reduction also does not occur directly to metallic iron; instead, $Fe_2O_3$ first converts into FeO, which was not initially present in the red mud, similar to carbothermal reduction.

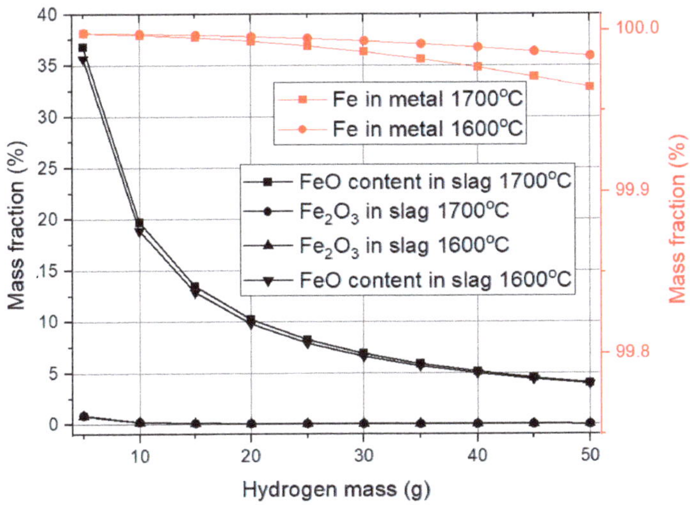


**Figure 6.** Mass fraction dependence on the hydrogen mass.

As hydrogen mass increases further, FeO content in the slag continues to decrease, reaching approximately 4% at 50 g of hydrogen. Beyond 25 g, the rate of decrease slows down, suggesting that most reducible iron oxides have already been converted into metallic iron. The iron content in the metal phase shows a slight decline as hydrogen mass increases, likely due to minor alloying with elements from the slag. However, this effect remains insignificant, as iron purity stays above 99.9%, demonstrating that the reduction process is highly effective in producing pure metallic iron.

Temperature variation from 1600 °C to 1700 °C does not have a major impact on the reduction process, except in the final iron content. At higher hydrogen masses, iron purity is slightly lower at 1700 °C than at 1600 °C, but the difference is negligible. An important observation is that hydrogen reduction produces significantly cleaner iron at all hydrogen mass levels compared to carbon reduction, highlighting its effectiveness in minimizing impurity incorporation. However, the efficiency of iron oxide reduction is slightly better for carbothermic reduction.

### 3.3. Carbothermic Reduction in an Electric Arc Furnace

In Figure 7, thermodynamic calculations of carbothermic and hydrogen reduction processes are compared to the experimentally obtained results.

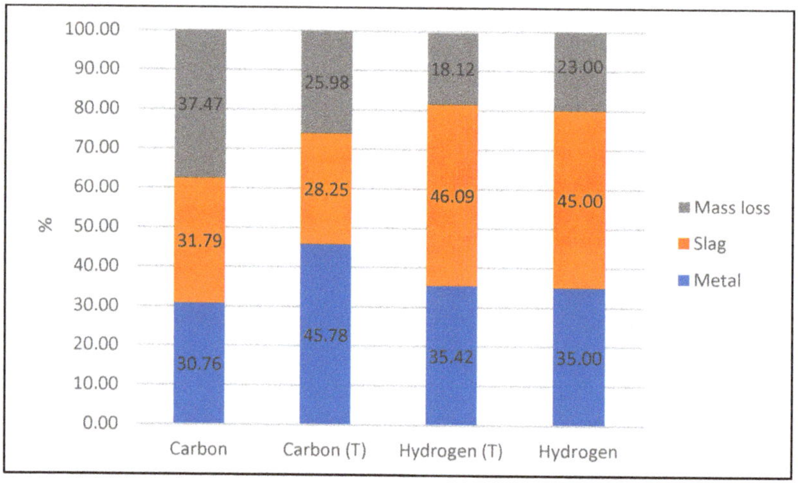


**Figure 7.** Thermodynamic calculation of the carbon and hydrogen reduction of red mud (T) and the experimentally obtained results.

Figure 7 shows the results comparison of the theoretically obtained results with the experimental results. It can be seen that there are slight variations, especially in carbothermic reduction, while for the hydrogen reduction process, theoretical and real experimental data are similar. Mass loss can be attributed to the reduction of oxygen from iron oxide (up to 15%) and evaporation in a neutral atmosphere (up to 9%). Particles obtained after hydrogen and carbothermic reduction exhibited magnetic properties, which are crucial for the separation process. ICP is used for elemental analysis of the carbothermic-reduced slag, as shown in Figure 8.

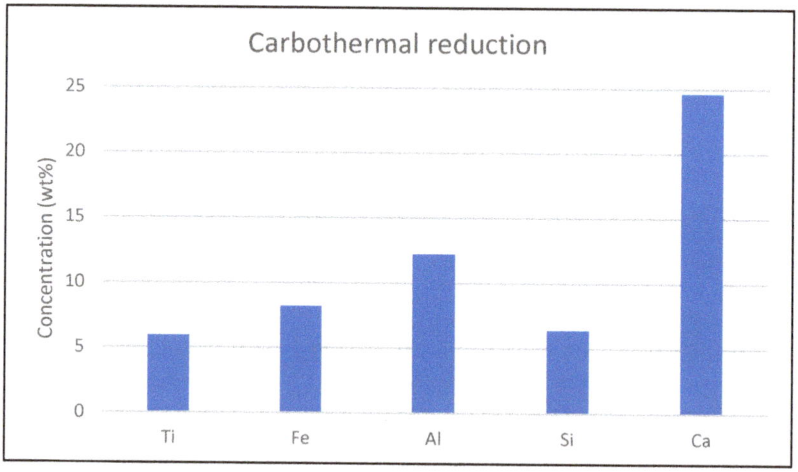


**Figure 8.** Elemental analysis of carbothermal-reduced slag at 1600 °C.

This Figure 8 shows the results of ICP analysis and composition (in weight percentage, wt%) of various elements, such as iron (Fe), titanium (Ti), silicon (Si), aluminum (Al), and calcium (Ca), in the slag after carbothermal reduction. The high concentration of calcium, approaching 30%, reflects the addition of CaO as a flux agent during the process. Compared to the original red mud, which contained a very low amount of calcium, the introduction of CaO significantly increases its concentration in the reduced slag. Aluminum also shows an increase in concentration, slightly above 10%. The concentration of iron is reduced to between 10% and, 5%, representing a successful reduction of iron oxides. Silicon and titanium are also significantly enriched in slag due to the inertness of this reducing agent. This figure highlights the impact of adding CaO as a flux agent, showing that it became dominant in the slag.

Reduced red mud powder samples were analyzed at room temperature by the X-ray powder diffraction technique. All obtained powders were identified using the ICDD database at https://www.icdd.com (accessed on 20 March 2025).

The X-ray diffraction (XRD) results of the carbothermic reduction of red mud show (Figure 9) that hematite, boehmite, diaspore, ilmenite, perovskite, quartz, and other minerals have been transformed. XRD patterns of red mud (Figure 1) show characteristic peaks attributed to these phases, with hematite being a major component due to its high iron content. Aluminum-bearing phases, boehmite, diaspore, and titanium-containing minerals (ilmenite and perovskite) are also present in red mud along with quartz, which is a very common silicate phase.

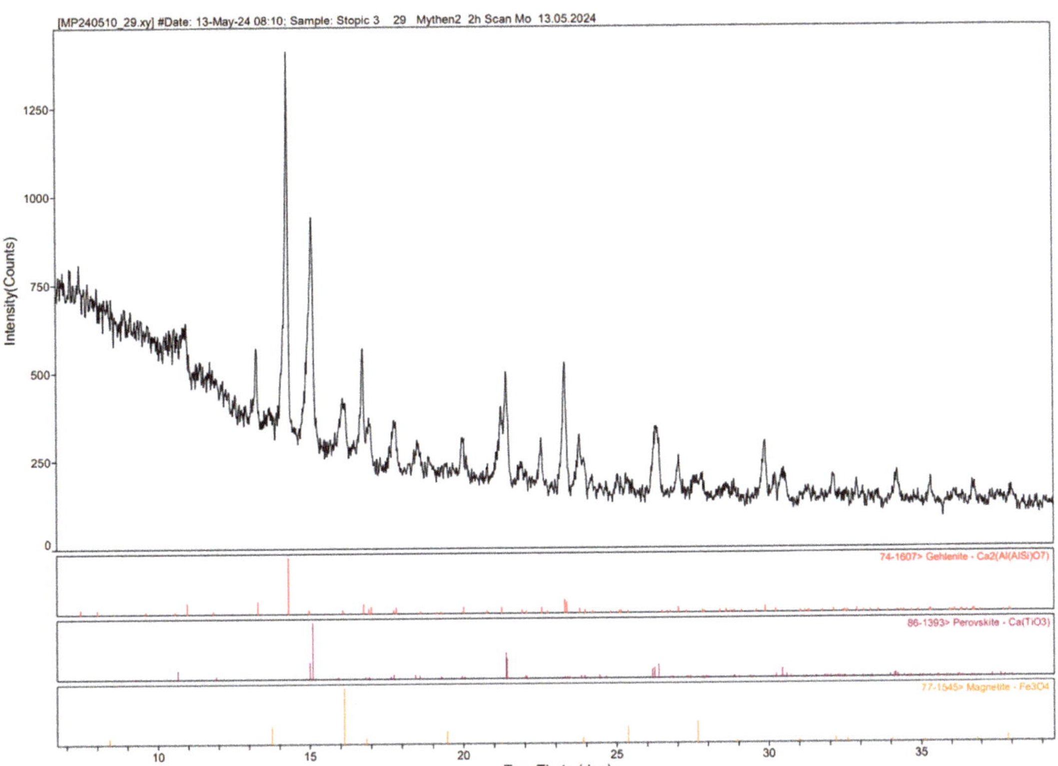


**Figure 9.** Diffractogram for slag obtained during red mud reduction at 1600 °C.

After the carbothermic reduction process, the XRD results are completely different. Hematite is no longer present because of its complete transformation. The reduction process produced two different phases: metallic (iron (Fe)) and slag. The XRD pattern of slag reveals the presence of perovskite, indicating that it remains largely unchanged, with the formation of gehlenite (calcium aluminum silicate) produced likely from the interaction of alumina with silica at high temperatures.

The absence of hematite peaks in the XRD pattern confirms its complete reduction, while the presence of magnetite suggests that some iron remained oxidized within the slag. The presence of perovskite indicates that titanium cannot be reduced, while the appearance of gehlenite reflects the complex transformations between the various oxides in red mud during the high-temperature exposure. Overall, the XRD results show the complete transformation of the red mud from a complex mixture of oxides and silicates to a system dominated by metallic iron and a slag phase, where perovskite and gehlenite are primary phases, with a minor amount of magnetite also present.

### *3.4. Hydrogen Reduction in the Rotary Kiln*

After the reduction process, the quartz tube has been removed from the rotary kiln, and the obtained powder has been analyzed.

The XRD analysis of the sample revealed three distinct phases: Fe, perovskite, and gehlenite (Figure 10). This result is from the reduction of red mud with hydrogen, where hematite, the main phase in red mud, has been reduced. The formation of new phases, such as perovskite and gehlenite, indicate successful transformation during the reduction process, highlighting the potential of this method for modifying red mud and creating valuable materials. A total of 24.65% of mass loss for red mud has been achieved during the reduction, including a 10% loss of ignition. This means that 14.65% has been attributed to the reduction process of removing oxygen from iron in the form of water vapor.

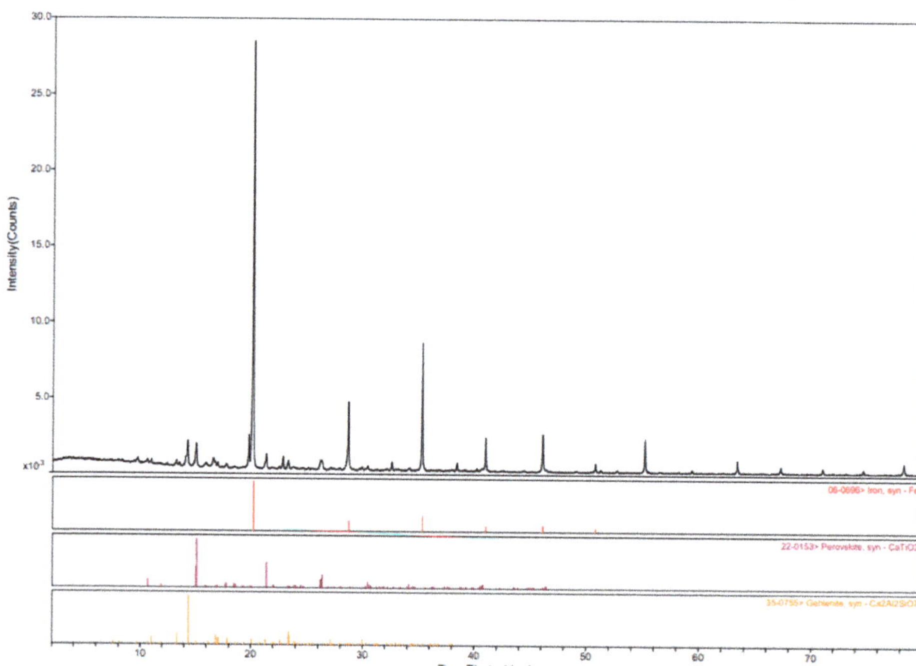


**Figure 10.** XRD analysis of hydrogen-reduced red mud.

### *3.5. Hydrogen Reduction in the Tubular Furnace*

The prepared powders obtained between 800 °C and 1000 °C have irregular, polygonal forms with particle sizes of more than 100 μm, as shown in Figure 11a,b.

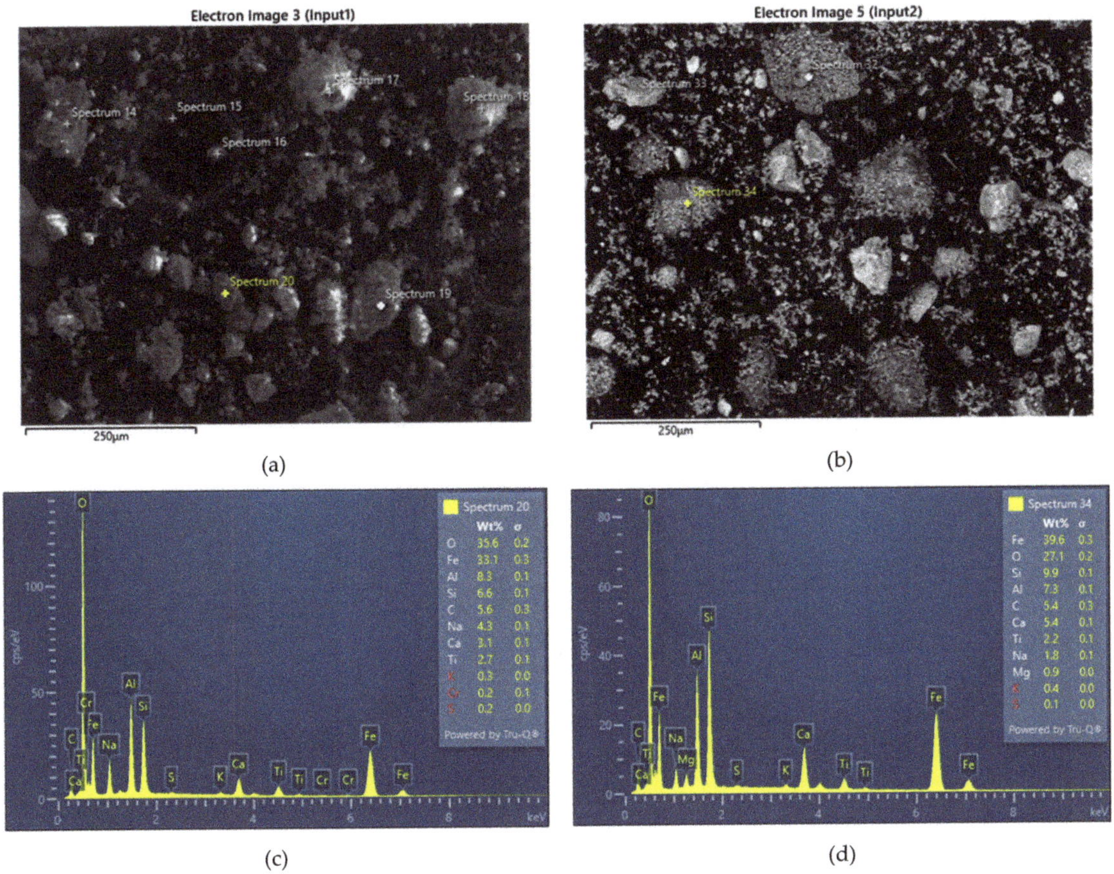


**Figure 11.** (**a**) SEM-Analysis of powder obtained by a hydrogen reduction of red mud at 800 °C; (**b**) SEM-Analysis of powder obtained by a hydrogen reduction of red mud at 1000 °C; (**c**) EDS-Analysis of powder obtained by a hydrogen reduction of red mud at 800 °C; (**d**) EDS-Analysis of powder obtained by a hydrogen reduction of red mud at 1000 °C.

An increase in temperature from 800 °C to 1000 °C increases the content of iron and decreases the content of oxygen, as shown in Figure 5. The morphological study of the reduced samples of iron oxides by $H_2$ shows agglomeration of the reaction product at temperatures between 800 and 1000 °C (Figure 11a,b).

The weight loss rate was calculated to indicate the degree of ilmenite hydrogen reduction. It was calculated as follows:

$$\alpha = (m_o - m_1/m_0) \times 100\% \tag{10}$$

where "$\alpha$" denotes the weight loss ratio and "$m_0$" and "$m_1$" represent the mass before and after the hydrogen reaction, respectively.

The calculated mass loss of the initial sample amounted between 21 and 25% (as shown in Table 4), which is according to the expected total theoretical value of the reduction process regarding the mass loss of oxygen from iron oxide, cobalt oxide, and nickel oxide through hydrogen reduction (max 15%) and evaporation in the neutral atmosphere (max 10%).

The obtained particles after hydrogen reduction have magnetic properties, which is very important for the following separation process.

**Table 4.** The experimental results for the reduction mass loss of red mud in a hydrogen atmosphere.

| Mass Loss (%) | 800 °C | 900 °C | 1000 °C |
|---|---|---|---|
| 60 min | 21 | 23 | 25 |

Diffractograms for three red mud samples reduced in a hydrogen atmosphere at 800, 900, and 1000 °C are shown in Figure 12. The identified phases include metallic iron, magnetite, perovskite, nepheline, and cancrinite.

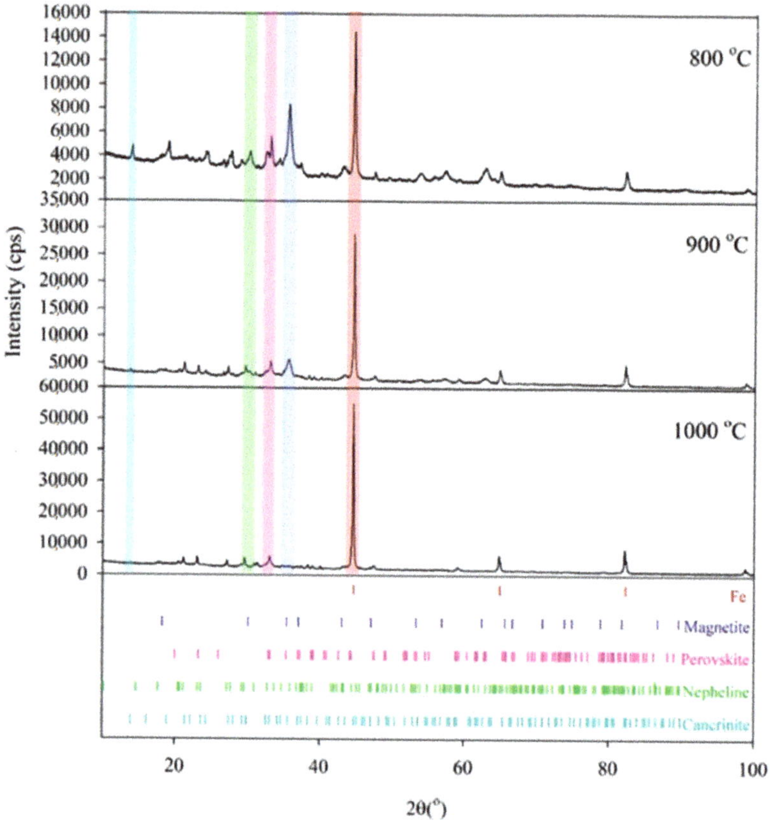


**Figure 12.** XRD analysis of the red mud reduced at 800, 900, and 1000 °C.

Metallic iron appears as a new phase not present in raw red mud, and it is characterized by a sharp diffraction peak that reflects the high crystallinity of iron. The intensities of iron diffraction peaks relative to those of other phases progressively increase with the reduction temperature, indicating an increase in iron abundance. At 900 °C, and especially at 1000 °C, metallic iron becomes the dominant phase.

Iron is partially present in magnetite, which is the most abundant phase in the sample reduced at 800 °C. The intensity of magnetite diffraction peaks and their content progressively decrease with increasing reduction temperature. A change in the position of the diffraction maxima compared to pure $Fe_3O_4$ may be attributed to the presence of Al, Ti, and Ca, which is consistent with the presence of these alloying elements, as determined by EDS analysis. A notable feature is the progressive increase in metallic iron content accompanied by a decrease in magnetite content with increasing reduction temperature. Magnetite is virtually absent in the sample reduced at 1000 °C, where only reduced elemental iron dominates.

Other non-ferrous phases identified include perovskite and nepheline, which are formed during the reduction process and are not present in the raw red mud, as well as residual cancrinite, which is present in raw red mud. Perovskite, the only Ti-bearing phase, appears to be stable across the examined temperature range (800–1000 °C), as inferred from XRD. Nepheline is the only formed silicate phase that is present in all products. The observed trend shows a decrease in residual cancrinite content accompanied by an increase in nepheline content with increasing reduction temperature, leading to the complete disappearance of cancrinite at the highest reduction temperature. This trend suggests the transformation of cancrinite into nepheline.

### *3.6. Hydrogen Plasma Reduction in the Electric Arc Furnace*

The plasma-thermal reduction was carried out in a reduction furnace at the Max Planck Institute for Sustainable Materials. It is performed using hydrogen plasma instead of molecular hydrogen, carbon, or carbon monoxide as the reducing agent.

After the reduction, the metal phase was separated from the slag using a magnet. Both phases were weighed, and the mass balance shown in Figure 13 was created. Since the metal phase was deposited in small droplets, it is possible that tiny metal droplets, invisible to the naked eye, remained in the slag phase. At the same time, small amounts of slag could adhere to the metal phase, leading to errors in the mass balance. Additionally, losses may occur during the handling of the samples.

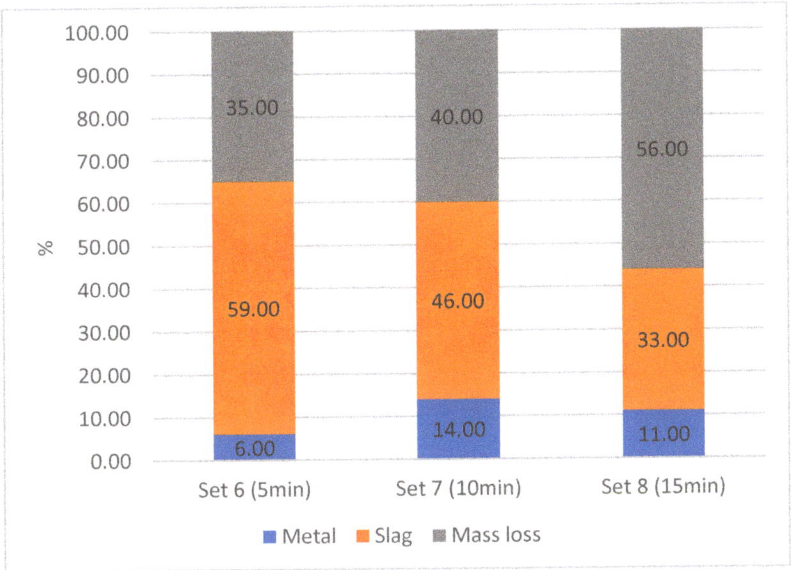


**Figure 13.** Mass balance of H-plasma reduction.

Figure 13 illustrates the mass balance observed during the plasma-thermal reduction process. Mass losses of up to 56% were achieved, a portion of which could be attributed to the reduction of oxides to metals, primarily aiming to produce a metallic iron phase and remove iron from the slag. However, analysis indicates that a significant part of the mass loss is not solely due to the reduction reaction but may also be caused by other factors, like moisture in the feed material, losses during handling, and dust or evaporation during the experiment. Figure 13 shows the highest amount of the metallic phase obtained with a 10 min reaction time, suggesting that it can be optimal for these parameters. The various results in mass loss for different experiments indicate potential dependence on unknown factors, like parameter fluctuations or inhomogeneous feed material.

Figure 14 shows the results of the XRD analysis of plasma-reduced red mud. It shows characteristic peaks for different compounds, and it can be seen that hematite is completely reduced to either magnetite and wustite or metallic iron. The analysis also detects iron titanium oxide ($FeTiO_3$) known as ilmenite, which is a primary mineral used for titanium extraction. As XRD analysis only identifies compounds in a crystalline structure, only silicon oxide (cristobalite) was identified, while the remaining components, like aluminum and calcium, detected by ICP, do not appear in the XRD analysis. This absence suggests that these elements are likely present in an amorphous phase, which cannot be detected by XRD. Agrawal [33] mentioned that gibbsite from red mud was dexydroxylated to alumina during hydrogen reduction at higher temperatures (750–900 °C). But, the transformation in hydrogen plasma conditions in a short time above 1800 °C shall be investigated in detail using quantitative XRD analysis.

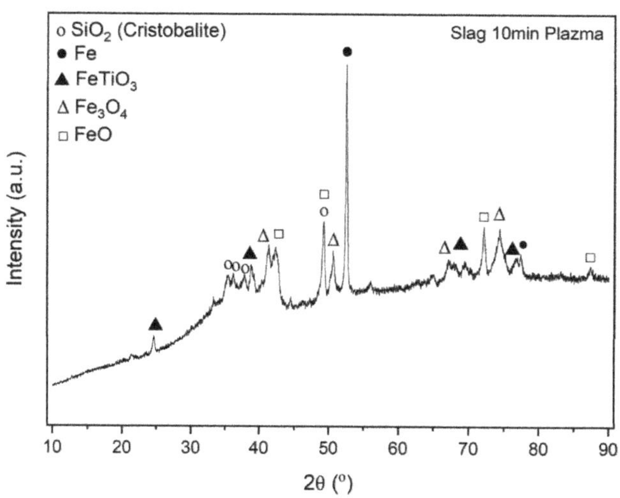


**Figure 14.** XRD analysis of a plasma thermally reduced red mud slag after 10 min of reduction time.

Figure 14 shows the limitations of XRD in making quantitative assessments, as it cannot determine all the elements that are present in the sample. Comparative analysis confirms that the reduction process has high significance in leaching efficiency, as shown in Table 5.

**Table 5.** Comparative analysis of the reduction process and leaching efficiency of solid residue (150°C, s/L: 1:10; 1 h, 500 rpm, pH < 0).

| Reduction | Rotary Kiln ($H_2$) | Electric Arc Furnace (H) | Electric Arc Furnace (C) [26] |
|---|---|---|---|
| Content of iron in solid residue (%) | 30 | 23 | 8 |
| Leaching efficiency at 150 °C (%) | 99 | 86 | 96 |

A reduction in all cases produces a product with magnetic properties. The obtained results confirmed that the reduction efficiency with carbon is highest, but it takes more time

(2 h) to produce titanium carbide and carbon dioxide, which is not environmentally friendly. The use of decarbonizing technology with plasma hydrogen in a short time (10 min) and with only 10% $H_2$ in the mixture can be optimized in order to produce metallic iron. But, the separation process of metallic iron by magnetic separation is most effective after hydrogen plasma reduction. The leaching efficiency is maximal for the solid residue after a reduction in the rotary furnace with hydrogen and in the electric arc furnace with carbon. The leaching efficiency of slag after hydrogen plasma reduction (86%) is higher than that in previous literature data, but it can be optimized through the reduction process in future research.

Maybe some pretreatment of red mud can increase reduction efficiency. Jena et al. [34] proposed a novel approach involving ultrasonication followed by water washing to pre-concentrate iron before reduction roasting of BR. Conventional reduction roasting of BR is more expensive compared to pre-concentrated BR, making this method potentially cost effective. The process consists of a two-stage ultrasonic treatment procedure. In the first stage, ultrasonication at 30 °C, with 40% solid content ($w/v$) for 30 min, improves the iron grade to 41.2 wt. (%) with 82% recovery. However, a second stage of ultrasonication is necessary to further upgrade the iron. Although the study optimized the ultrasonication time, temperature, solid content (%, $w/v$), and water washing parameters, an application in scale-up conditions shall be studied in detail using red mud from Alumina, Bosnia.

## 4. Conclusions

This study investigated the reduction of iron oxides from red mud using both carbothermal and hydrogen-based reduction methods, combining thermochemical modeling and experimental validation.

Thermochemical calculations provided insights into the reduction mechanisms of red mud. In the carbothermal reduction model, $Fe_2O_3$ was initially reduced to FeO before further transforming into metallic iron. The process was strongly influenced by carbon mass, with an optimal range between 10 g and 15 g to maximize metal recovery while minimizing impurity incorporation. Temperature had a moderate impact, with 1700 °C showing slightly improved metal yield. Similarly, the hydrogen reduction model indicated that $Fe_2O_3$ first converts into FeO before reducing to metallic iron, with 20 g of hydrogen identified as the optimal amount for effective reduction. Hydrogen reduction resulted in significantly purer iron compared to carbothermal reduction, with minimal impurity incorporation.

Experimental carbothermal reduction confirmed the thermochemical modeling results, with iron successfully extracted while impurities, such as silicon and titanium, remained in the slag. The addition of CaO as a flux agent significantly increased the calcium content in the slag. X-ray diffraction (XRD) analysis confirmed complete hematite reduction, resulting in metallic iron and a slag phase dominated by perovskite and gehlenite. Some magnetite remained in the slag, indicating an incomplete reduction of iron oxides.

Hydrogen-based reduction was explored using three different approaches: a tubular furnace, a rotary kiln, and hydrogen plasma reduction. In the tubular furnace, increasing the temperature from 800 °C to 1000 °C enhanced iron reduction, with iron purity increasing as oxygen content decreased. XRD confirmed the transformation of iron oxides into metallic iron, with residual magnetite decreasing at higher temperatures. The presence of perovskite and nepheline suggested interactions between alumina, silica, and titanium-bearing phases. Reduction in a rotary kiln resulted in a 24.65% mass loss, primarily due to oxygen removal in the form of water vapor. The process effectively converted hematite into metallic iron, with the presence of perovskite and gehlenite indicating phase transformations during reduction. Hydrogen plasma reduction achieved the highest mass loss, up to 56%, attributed to the reduction of oxides and process-related factors, such as evaporation and sample handling.

The iron content in the slag significantly decreased, confirming the efficiency of plasma reduction for iron extraction. XRD analysis identified iron titanium oxide (ilmenite) and suggested the presence of amorphous aluminum and calcium phases.

Both carbothermal and hydrogen-based reduction methods demonstrated effective iron recovery from red mud. Hydrogen reduction methods produced higher-purity iron, with the plasma process achieving the most significant reduction. The carbothermal reduction was more efficient in terms of overall metal yield but introduced more impurities. Among hydrogen-based techniques, plasma reduction achieved the highest iron extraction efficiency, while reduction in a tubular furnace and a rotary kiln provided scalable alternatives with good iron purity. These findings highlight the potential for hydrogen-based reduction as a cleaner alternative for iron recovery from red mud while also emphasizing the importance of optimizing parameters to balance yield and purity.

**Author Contributions:** Conceptualization, S.S. and D.K.; methodology, S.S.; software, A.M. and R.S.; validation, S.S., D.K. and I.R.S.F.; formal analysis, R.S. and I.R.S.F.; investigation, I.R.S.F., M.P. and D.K.; resources, S.S.; data curation, S.S.; writing—original draft preparation, D.K. and S.S.; writing—review and editing, I.R.S.F., M.P. and A.M.; visualization, D.K. and A.M.; supervision, B.F. and M.P.; project administration, S.S.; funding acquisition, S.S. and M.P. All authors have read and agreed to the published version of the manuscript.

**Funding:** This research was funded by the European Commission, grant number 101135077 (EURO-TITAN project).

**Data Availability Statement:** Data are contained within the article.

**Conflicts of Interest:** The authors declare no conflicts of interest.

## References

1. Spreitzer, D.; Schenk, J. Reduction of Iron Oxides with Hydrogen—A Review. *Steel Res. Int.* **2019**, *90*, 1900108. [CrossRef]
2. Tokumitsu, K. Reduction of Metal Oxides by Mechanical Alloying Method. *Solid State Ion.* **1997**, *101–103*, 25–31. [CrossRef]
3. Rukini, A.; Rhamdhani, M.A.; Brooks, G.A.; Van den Bulck, A. Metals Production and Metal Oxides Reduction Using Hydrogen: A Review. *J. Sustain. Metall.* **2022**, *8*, 1–24. [CrossRef]
4. The Thermite Reaction Between Aluminium and Iron (III) Oxide | Experiment | RSC Education. Available online: https://edu.rsc.org/experiments/the-thermite-reaction-between-aluminium-and-ironiii-oxide/724.article (accessed on 23 July 2024).
5. Cheema, H.A.; Ilyas, S.; Farhan, M.; Yang, J.-H.; Kim, H. Extraction of Fe and Al from Red Mud Using Carbothermic Reduction Followed by Water Leaching. *J. Ind. Eng. Chem.* **2024**, *136*, 201–210. [CrossRef]
6. Blast Furnace Process—Tec-Science. Available online: https://www.tec-science.com/material-science/steel-making/blast-furnace-process/ (accessed on 23 July 2024).
7. Habibi, H.; Piruzian, D.; Shakibania, S.; Pourkarimi, Z.; Mokmeli, M. The Effect of Carbothermal Reduction on the Physical and Chemical Separation of the Red Mud Components. *Miner. Eng.* **2021**, *173*, 107216. [CrossRef]
8. Zinoveev, D.; Pasechnik, L.; Grudinsky, P.; Yurtaeva, A.; Dyubanov, V. Kinetics and Mechanism of Red Mud Carbothermic Reduction and Reduced Iron Grain Growth: An Influence of Sodium Sulfate. *Crystals* **2023**, *13*, 1. [CrossRef]
9. Lackner, A.; Molinari, T.; Paschen, P. A thermogravimetric study on the hydrogen reduction of tungsten oxides. *Scand. J. Metall.* **1996**, *25*, 115–121.
10. European Commision. A Roadmap for Moving to a Competitive Low-Carbon Economy in 2050—European Environment Agency. Available online: https://www.eea.europa.eu/policy-documents/com-2011-112-a-roadmap (accessed on 21 July 2024).
11. Cavaliere, P. Clean Ironmaking and Steelmaking Processes: Efficient Technologies for Greenhouse Emissions Abatement. In *Clean Ironmaking and Steelmaking Processes*; Springer: Cham, Switzerland, 2019; pp. 1–37.
12. Otto, A.; Robinius, M.; Grube, T.; Schiebahn, S.; Praktiknjo, A.; Stolten, D. Power-to-Steel: Reducing $CO_2$ through the Integration of Renewable Energy and Hydrogen into the German Steel Industry. *Energies* **2017**, *10*, 451. [CrossRef]
13. Souza Filho, I.R.; Ma, Y.; Kulse, M.; Ponge, D.; Gault, B.; Springer, H.; Raabe, D. Sustainable Steel through Hydrogen Plasma Reduction of Iron Ore: Process, Kinetics, Microstructure, Chemistry. *Acta Mater.* **2021**, *213*, 116971. [CrossRef]
14. Raabe, D. The Materials Science behind Sustainable Metals and Alloys. *Chem. Rev.* **2023**, *123*, 2436–2608. [CrossRef]
15. Souza Filho, I.R.; Springer, H.; Ma, Y.; Mahajan, A.; da Silva, C.C.; Kulse, M.; Raabe, D. Green Steel at Its Crossroads: Hybrid Hydrogen-Based Reduction of Iron Ores. *J. Clean. Prod.* **2022**, *340*, 130805. [CrossRef]

16. Cavaliere, P. Hydrogen Plasma Reduction. In *Hydrogen Assisted Direct Reduction of Iron Oxides*; Springer: Cham, Switzerland, 2022; pp. 321–338.
17. Tang, W.; Khavarian, M.; Yousefi, A. Red Mud. Sustainable Concrete Made with Ashes and Dust from Different Sources. *Mater. Prop. Appl.* **2022**, 577–606.
18. Liu, Y.; Naidu, R. Hidden Values in Bauxite Residue (Red Mud): Recovery of Metals. *Waste Manag.* **2014**, *34*, 2662–2673. [CrossRef] [PubMed]
19. Qi, Y. The Neutralization and Recycling of Red Mud-a Review. *J. Phys. Conf. Ser.* **2021**, *1759*, 12004. [CrossRef]
20. Verma, A.S.; Suri, N.M.; Kant, S. Applications of Bauxite Residue: A Mini-Review. *Waste Manag. Res.* **2017**, *35*, 999–1012. [CrossRef]
21. Tanvar, H.; Merino, K.; Mishra, B. Extraction of High Purity Magnetite from Bauxite Residue. *Miner. Eng.* **2023**, *199*, 108131. [CrossRef]
22. Rai, S.; Bahadure, S.; Chaddha, M.J.; Agnihotri, A. Disposal Practices and Utilization of Red Mud (Bauxite Residue): A Review in Indian Context and Abroad. *J. Sustain. Metall.* **2020**, *6*, 1–8. [CrossRef]
23. Winkler, D.; Bidló, A.; Bolodár-Varga, B.; Erdő, Á.; Horváth, A. Long-Term Ecological Effects of the Red Mud Disaster in Hungary: Regeneration of Red Mud Flooded Areas in a Contaminated Industrial Region. *Sci. Total Environ.* **2018**, *644*, 1292–1303. [CrossRef]
24. FactSage Software and Databases. Available online: https://factsage.com/ (accessed on 6 March 2025).
25. Stopic, S.; Kostić, D.; Schneider, R.; Sievers, M.; Wegmann, F.; Emil Kaya, E.; Perušić, M.; Friedrich, B. Recovery of Titanium from Red Mud Using Carbothermic Reduction and High Pressure Leaching of the Slag in an Autoclave. *Minerals* **2024**, *14*, 1151. [CrossRef]
26. Menad, N.; Kanari, N. Recovery of high grade iron compounds from LD slag by enhanced magnetic separation techniques. *Int. J. Miner. Process.* **2014**, *126*, 1–9. [CrossRef]
27. Pineau, A.; Kanari, N.; Gaballah, I. Kinetics of Reduction of Iron Oxides by $H_2$: Part I: Low Temperature Reduction of Hematite. *Thermochim. Acta* **2006**, *447*, 89–100. [CrossRef]
28. Pilla, G.; Kapelari, S.V.; Hertel, T.; Blanpain, B.; Pontikes, Y. Hydrogen Reduction of Bauxite Residue and Selective Metal Recovery. *Mater. Today Proc.* **2022**, *57*, 705–710. [CrossRef]
29. Lazou, A.; van der Eijk, C.; Balomenos, E.; Kolbeinsen, L.; Safarian, J. On the Direct Reduction Phenomena of Bauxite Ore Using $H_2$ Gas in a Fixed Bed Reactor. *J. Sustain. Metall.* **2020**, *6*, 227–238. [CrossRef]
30. Taskinen, P.; Hytönen, P.; Tikkanen, H. On the Reduction of Tungsten Oxides, II. *Scand. J. Metall.* **1977**, *6*, 228–232.
31. Stopic, S.; Nikolić, N.; Životić, D.; Krstić, J.; Rosić, M.; Damjanović, V.; Friedrich, B. Hydrogen Metallurgy for a Treatment of Bauxite Residues-First Results on the EURO-TITAN Project. In Proceedings of the XXV YuCorr; Ministry of Science, Technological Development and Innovation of the Republic of Serbia, Divčibare, Serbia, 29 May 2024; pp. 11–20.
32. Balomenos, E.; Gianopoulou, I.; Panias, D.; Paspaliaris, I. ENEXAL: Novel technologies for enhanced energy and exergy efficiencies in primary aluminium production industry. *Metall. Mater. Eng.* **2009**, *15*, 203–217.
33. Agrawal, S.; Dhawan, N. Hydrogen Reduction of Red Mud for Extraction of Metallic Values. *J. Sustain. Metall.* **2023**, *9*, 386–397. [CrossRef]
34. Jena, S.; Nayak, A.; Das, S. Ultrasonic-assisted pre-concentration of iron values in bauxite residue suitable for reduction-roasting purpose. *Chem. Eng. Commun.* **2025**, *212*, 4. [CrossRef]

MDPI AG
Grosspeteranlage 5
4052 Basel
Switzerland
Tel.: +41 61 683 77 34

*Minerals* Editorial Office
E-mail: minerals@mdpi.com
www.mdpi.com/journal/minerals

www.ingramcontent.com/pod-product-compliance
Lightning Source LLC
Chambersburg PA
CBHW062049220926
51850CB00010B/2709
* 9 7 8 3 7 2 5 8 5 6 2 9 9 *